SpringerBriefs in Water Science and Technology

AF844149

SpringerBriefs in Water Science and Technology present concise summaries of cutting-edge research and practical applications. The series focuses on interdisciplinary research bridging between science, engineering applications and management aspects of water. Featuring compact volumes of 50 to 125 pages (approx. 20,000–70,000 words), the series covers a wide range of content from professional to academic such as:

- Literature reviews
- In-depth case studies
- Bridges between new research results
- Snapshots of hot and/or emerging topics

Topics covered are for example the movement, distribution and quality of freshwater; water resources; the quality and pollution of water and its influence on health; and the water industry including drinking water, wastewater, and desalination services and technologies.

Both solicited and unsolicited manuscripts are considered for publication in this series.

Andrea Guerrini

Private Investments in the European Water Sector

A Country Overview Based on Utilities' Performance

Springer

Andrea Guerrini
Department of Management
University of Verona
Verona, Italy

ISSN 2194-7244 ISSN 2194-7252 (electronic)
SpringerBriefs in Water Science and Technology
ISBN 978-3-032-24734-6 ISBN 978-3-032-24735-3 (eBook)
https://doi.org/10.1007/978-3-032-24735-3

© The Editor(s) (if applicable) and The Author(s), under exclusive license to Springer Nature Switzerland AG 2026

This work is subject to copyright. All rights are solely and exclusively licensed by the Publisher, whether the whole or part of the material is concerned, specifically the rights of translation, reprinting, reuse of illustrations, recitation, broadcasting, reproduction on microfilms or in any other physical way, and transmission or information storage and retrieval, electronic adaptation, computer software, or by similar or dissimilar methodology now known or hereafter developed.
The use of general descriptive names, registered names, trademarks, service marks, etc. in this publication does not imply, even in the absence of a specific statement, that such names are exempt from the relevant protective laws and regulations and therefore free for general use.
The publisher, the authors and the editors are safe to assume that the advice and information in this book are believed to be true and accurate at the date of publication. Neither the publisher nor the authors or the editors give a warranty, expressed or implied, with respect to the material contained herein or for any errors or omissions that may have been made. The publisher remains neutral with regard to jurisdictional claims in published maps and institutional affiliations.

This Springer imprint is published by the registered company Springer Nature Switzerland AG
The registered company address is: Gewerbestrasse 11, 6330 Cham, Switzerland

If disposing of this product, please recycle the paper.

Competing Interests The author has no competing interests to declare that are relevant to the content of this manuscript.

Contents

About the Author

Andrea Guerrini is a Professor in the Department of Management at the University of Verona. He has extensive experience in the field of management and the regulation of local public services, with a particular focus on the water, energy, and waste sectors. He served as a Member of the Italian Regulatory Authority for Energy, Networks and Environment (ARERA), President of the Association of European Water Regulators (WAREG), and a Member of the Bureau of the OECD Network of Economic Regulators (NER) from 2018 to 2025.

He has been President of a major water utility company operating in Tuscany and has also worked as a consultant for public bodies and companies in the areas of tariff regulation and performance management. He is the author of numerous publications in leading international scientific journals, for some of which he has also served as Editor (including *Water Resources Management*, *Sustainability*, *Water*, *Journal of Cleaner Production*, and *Water Policy*). He has participated in several research projects funded by the European Union (Life and H2020) and other institutions.

Chapter 1
Introduction of the Report

The strategic importance of resilient water and wastewater infrastructure systems has increased due to climate change. Consequently, stakeholders such as politicians, government officials, policy makers, institutional investors, and private investors have directed their attention towards this sector. The political relevance of the sector is reflected in legislative initiative, as the recent Drinking Water Directive (DWD) and the Urban Wastewater Directive enacted in Europe (UWWTD).

However, despite having a common set of rules, European water utilities operate within varying institutional frameworks, resulting in different performances that affect their attractiveness to private investors. The substantial amount of information published about the water sector in recent years is insufficient for accurately describing the country-risk profile for water investments, considering water governance, regulatory settings, and sector-specific features.

This report caters to the interests of a diverse group of stakeholders, primarily Water Regulatory Authorities (WRAs) and investors. It examines the performance of a selected panel of privately owned utilities in the European water sector with the primary aim of assessing the industry's appeal to private investors. It is based on the analysis of rate of returns and of their key drivers such as profitability, turnover, and financial leverage. Additionally, the report outlines the risk profiles of the selected countries. The observed measures also help to identify the impact of certain institutional variables, including country characteristics and water governance structures, by clustering countries based on the presence of an independent national regulator. Furthermore, the scope of operation is examined, differentiating between companies that offer only water services and those that operate under an integrated business model encompassing both water and wastewater management.

The selected panel comprises 32 entities operating in 10 countries, serving approximately 50.5 mn inhabitants, which represents around 10% of the combined population of Europe, England, and Wales. Given that a significant portion of the

© The Author(s), under exclusive license to Springer Nature Switzerland AG 2026

A. Guerrini, *Private Investments in the European Water Sector*, SpringerBriefs in Water Science and Technology, https://doi.org/10.1007/978-3-032-24735-3_1

European population is served by municipalities or state-owned companies, this rate can be considered fairly representative.

The designed panel was monitored using a list of indicators collected for the period 2018-2023, utilizing ORBIS, a dataset providing financial information for legal entities globally. Performance evaluation was conducted using the Return on Equity (ROE) pyramid.

This report is structured as follows: Chap. 2 provides an overview of the European water sector, highlighting its characteristics, water governance issues, and challenges. Chapter 3 begins with a methodological section followed by a description of the data collected and main results. Finally, Chap. 4 discusses the findings and introduces their implications.

Chapter 2
The Water Sector in Europe

2.1 The Sector in Figures

The water sector provides three essential services to households and non-domestic consumers: drinking water distribution, wastewater collection, and wastewater treatment.

In EU countries the sector is featured by 27,885 companies, which generate a turnover of 88 bn of euros, and employ approximately 485,000 persons (Annex I) for managing and maintaining a complex infrastructural network.

2.1.1 The Drinking Water Segment

According to the most recent data of Eurostat, in Europe the largest part of water resource is abstracted from surface waters: 75% of total water abstraction is from rivers and reservoirs, and 25% from groundwater. Agriculture is the sector using the largest amount of water (58%), followed by energy cooling (18%), mining (11%) and households (10%).[1] An increasing trend in water abstraction has featured the southern Europe, with a + 30% for electricity cooling, and a + 9% for agriculture since 1990. The Water consumption for agriculture has increased significantly in Turkey (70%) and Greece (11%), whereas electricity cooling has increased its consumption in the Netherlands (23%) and Slovenia (18%).

The average water resources available for multiple usage are about 4–5 thousand m^3 per inhabitant. In Croatia the inhabitant's share reaches around 30 thousand m^3, while other countries are at risk of "water stress", as Poland, Czechia, Cyprus and

[1] 1 https://water.europa.eu/freshwater/europe-freshwater.

© The Author(s), under exclusive license to Springer Nature Switzerland AG 2026

A. Guerrini, *Private Investments in the European Water Sector*, SpringerBriefs in Water Science and Technology, https://doi.org/10.1007/978-3-032-24735-3_2

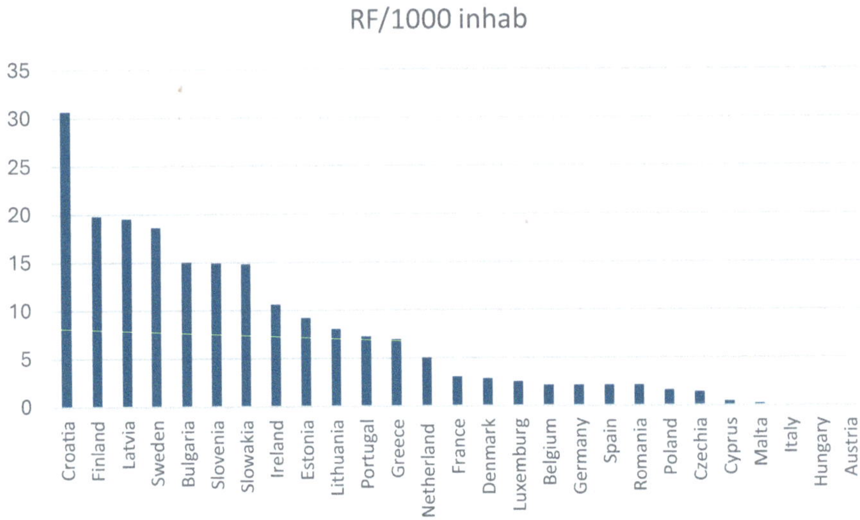

Fig. 2.1 Fresh water resource per inhabitant (*Eurostat*)

Malta, with an annual water resource below 1700 m^3 per inhabitant (Eurostat) (Fig. 2.1).

The Water exploitation index plus (WEI+) measures the percentage of renewable freshwater resources consumed at river basin or subunit levels. The water consumed is given by the differences between water abstracted and the volume of water returned to the environment by economic sectors before or after use. WPI+ signals water scarcity conditions: values above 20% indicate a stress, while values above 40% signal severe water stress.

Figure 2.2 shows that water scarcity conditions affected 34% of the population and 40% of the EU land area in 2022.

Moving from a total consumption of 243,000 mn m^3, approximately 50 bn m^3 of freshwater was extracted annually for public supply, serving about 500 mn of European citizens; this translates to an average daily usage rate of about 245 liters per person (Fig. 2.3; Magagna et al., 2019).

In terms of physical assets, the length of the drinking water network per connected inhabitant ranges from 4.35 mt per inhabitant in Romania to 19.57 mt in Finland. The total length of the drinking water network spans 4.3 mn km of pipes, while the wastewater network amounts to 3.2 mn km.

Water losses is an index which could effectively show the status of robustness of infrastructures: the data of Eureau report are reported in Fig. 2.4 and signal that several countries are far beyond the European average of approximately 23%,[2] which is kept quite low by those countries which are not part of the network, as

[2] www.environment.ec.europa.eu.

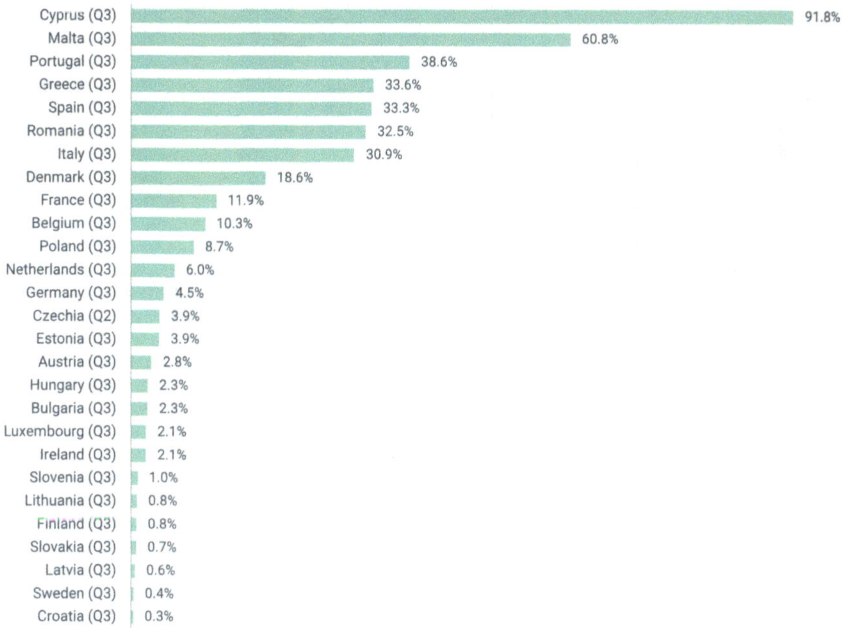

Fig. 2.2 WEI+ in European countries in 2022. (*EEA*)

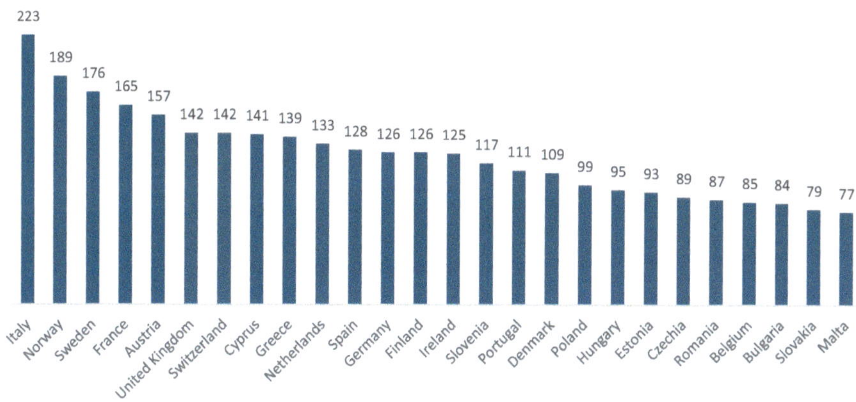

Fig. 2.3 Average residential drinking water daily consumption per capita in 2021 (*Statista on Eurostat data*)

Denmark, France and Germany, achieving good performance, with losses lower than 20% (EurEau 2021) (Fig. 2.5).

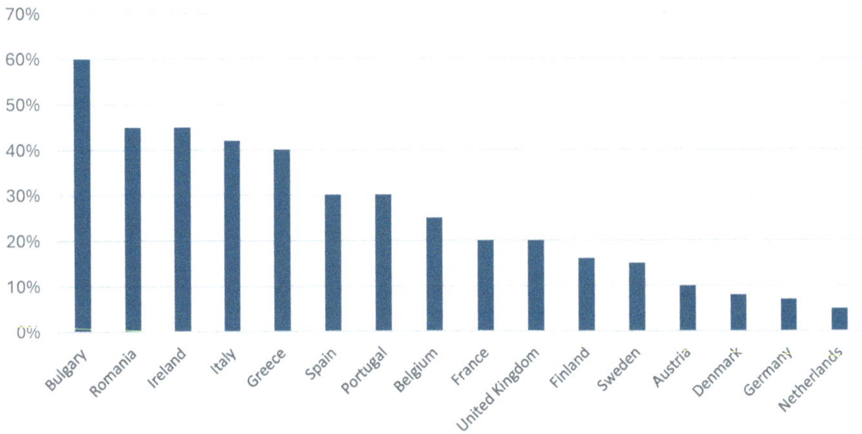

Fig. 2.4 Water losses in European countries (*Eureau data*)

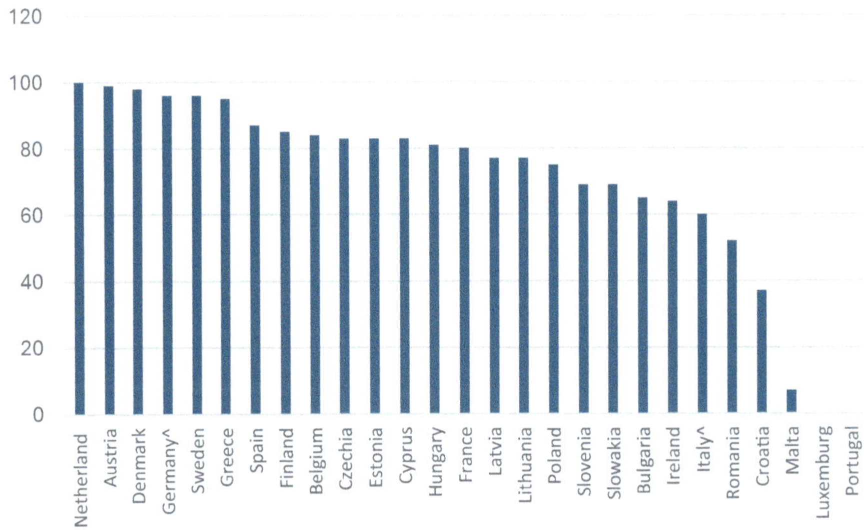

Fig. 2.5 Rate of population connected to a secondary Wastewater treatment plant in 2020 (Eurostat). ^ data of 2015

2.1.2 The Wastewater Segment

According to EurEau (2021), in Europe 23.6% of the collected wastewater is treated for organic matter removal, 63.7% for nutrient removal, and 11.3% for more stringent requirements than those specified by the Urban Wastewater Treatment Directive (UWWTD).

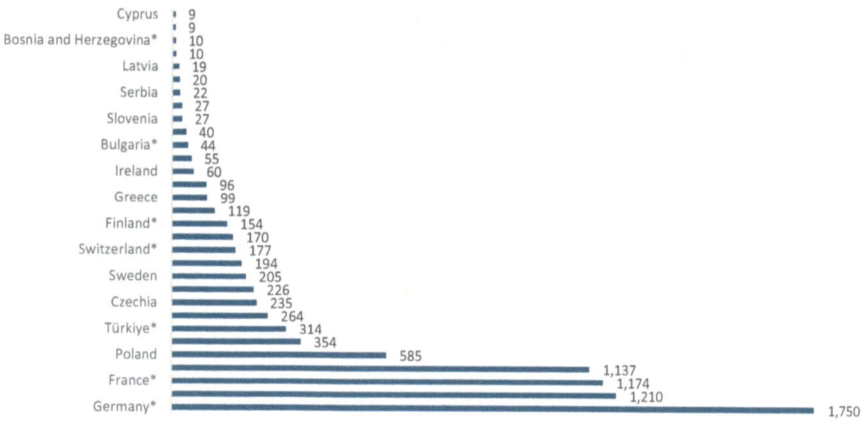

Fig. 2.6 Sewage sludge generation in Europe 2021 in metric tons (*Statista on Eurostat data*)

In half of EU member states, approximately the 80% of the population is connected to a secondary wastewater treatment plant, which is the facility required to provide a minimal acceptable level of environmental protection. The rate rises to 95% in Denmark, Germany, Greece, the Netherlands, Austria and Sweden; while, at the other extreme of the range, in Malta and Croatia the rate doesn't achieve the 50%. A high rate of connection is associated to a relevant production of sludge through the process of wastewater treatment (Fig. 2.6). EU generated over 6 tonnes of wastewater sludge in 2018, which are mainly destined in agriculture, for soil amelioration (36%), the 14.8% in incinerated and a 5.8% is disposed in landfill. The rest is differently disposed or no information is available on how the process of sludge treatment has been closed.

Despite the steps forward realized during the implementation of the UWWTD 91/271/EEC, further efforts are still needed to provide:

- the collection of additional 9.68 P.E. of wastewater;
- the biological treatment to approximately 44 mn P.E.;
- the biological treatment with nitrogen and/or phosphorus removal to 28 mn P.E.

According to the European Commission, the compliance of the total generated load in the EU countries is 82%,[3] with stable distance from targets for Collection and for Biological treatment during the period 2014–2018; and with a gradual reduction for Biological treatment and for nitrogen and/or phosphorous removal (Fig. 2.7).

The Fig. 2.8 reports the areas in compliance with the UWWTD requirements: the blue dot identifies a system of wastewater collection and treatment that meets legal requirements; the red dot shows a non-compliance status; while the gray identifies those systems not assessed. The level of compliance among different States is strictly related to the amount of investments per citizen per year for new collecting

[3] https://water.europa.eu/freshwater/countries/uwwt/european-union.

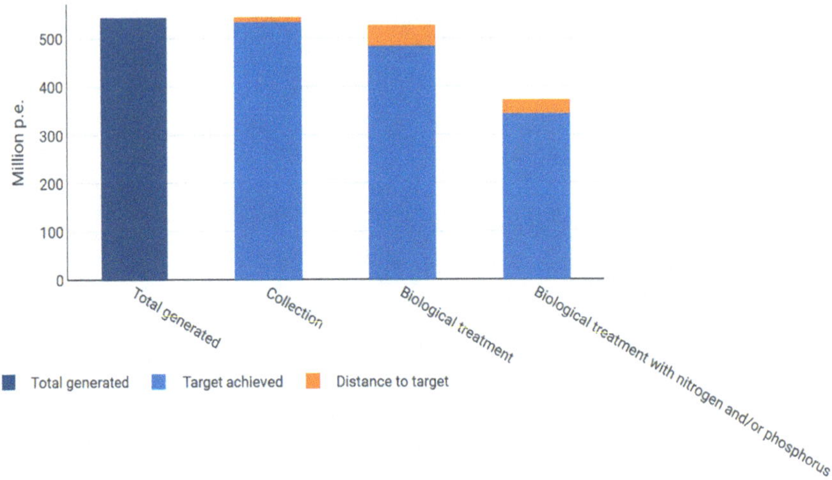

Fig. 2.7 Amount of urban wastewater not complying with the requirements of the UWWTD (https://water.europa.eu/freshwater/countries/uwwt/european-union)

and treatment infrastructures and for their refurbishment: with an average of 16.9 € at EU level, Sweden, Denmark, Luxemburg, Netherland and Finland realize annual investments for more than 50 € and achieve 207 € for Sweden.

2.1.3 *Toward a "New Water Paradigm"*

Beside these typical core activities, in recent years the role of water utilities was extended to include additional services provided directly to consumers or to the environment:

- recharging groundwater aquifers;
- producing treated wastewater for irrigation and industrial purposes;
- generating energy through water networks using hydropower plants;
- producing biogas and biomethane through proper sludge valorization.

Recently water utilities have facing significant challenges due to climate change and rapid population growth, which are addressed through innovative solutions, such as desalination and water reuse, to overcome water scarcity issues affecting conventional water sources.

Actually, the water sector has been transformed in a circular productive process, according to a "new water paradigm"[4] to deliver high quality drinking water and an

[4] https://sdgs.un.org/partnerships/water-climate-healing-new-water-paradigm.

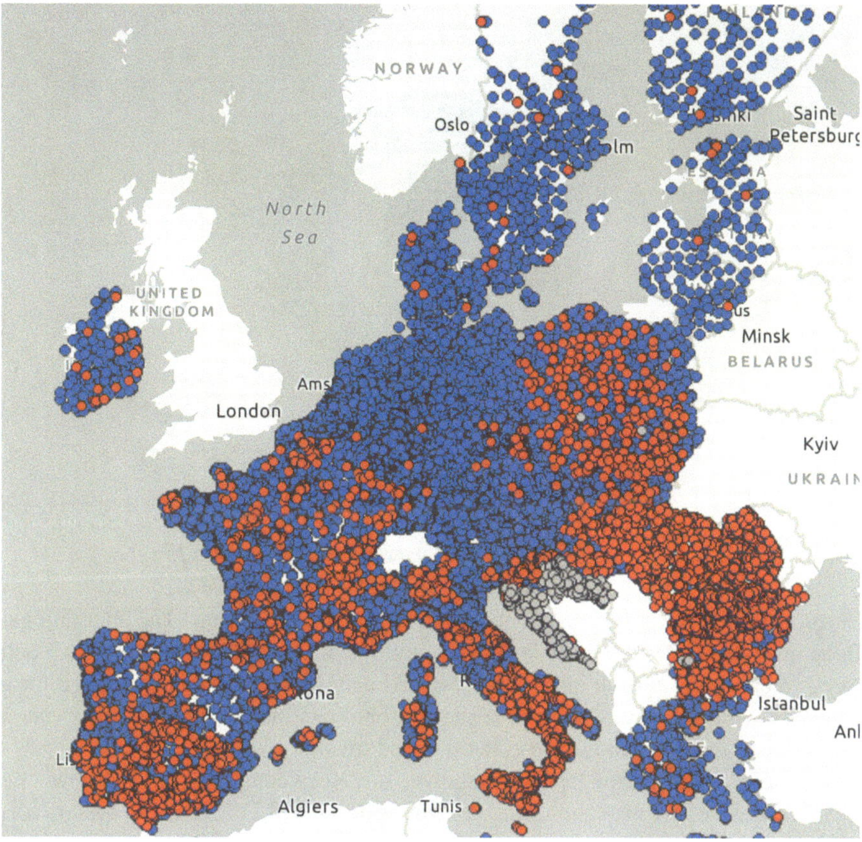

Fig. 2.8 Urban compliance status. (https://water.europa.eu/freshwater/countries/uwwt/european-union)

effective treatment of sewerage, accompanied by an extended range of byproducts as shown in Fig. 2.9.

According to the OECD (2020), the current water and wastewater services expenditure across the EU member states is estimated at 100 bn of euros. The annual per capita expenditure varies among member states, with some spending over 250 €, while others spend less than 100 €. EurEau (2021) estimates that investments in infrastructure amount to approximately 45 bn of euros, corresponding to 82 € per inhabitant per year. To address the upcoming challenges in the water sector the OECD report suggests a total cumulative additional expenditure of 289 bn of euros by 2030, thus implying an increase of over 20% in current annual investments.

The tariff system serves as the primary source of financing for current expenditures, accounting for approximately 70%, whereas public resources cover the remaining part of the financial need.

In terms of water prices, Europe is featured by wide differences among countries. Figure 2.7 represents the price in USD/m^3 of water services for different

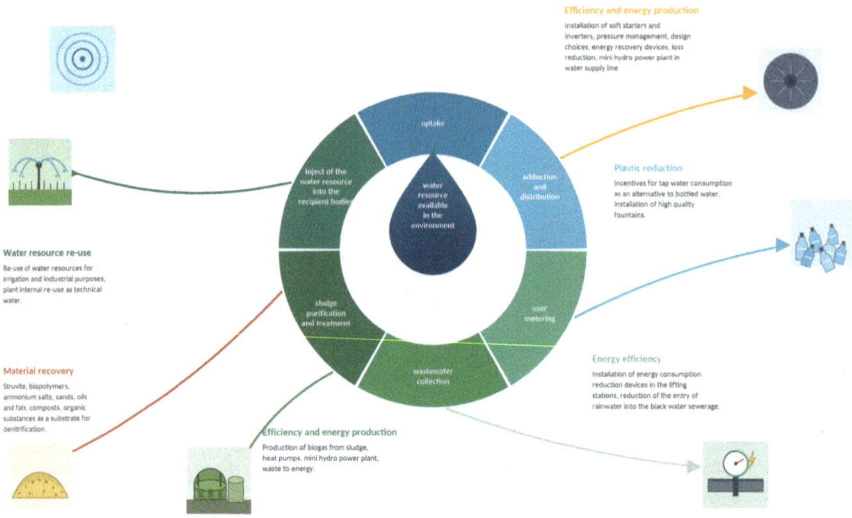

Fig. 2.9 The circularity of the integrated water sector (Guerrini and Manca 2020)

European cities: the price is based on the latest effective tariff from "The International Benchmarking Network for Water and Sanitation Utilities" (IBNET) database. Each city is represented by one utility provider and its respective rate is calculated for a consumption up to 15 m^3 per month per household and is referred to the provision of drinking tap water, and to the collection and treatment of wastewater.

Of the selected European cities shown, the Norwegian capital Oslo had the most expensive tap water in 2021, averaging 6.69 U.S.D. per cubic meter; followed by Germany's manufacturing hub Stuttgart, where 1000 liters of tap water cost 5.67 U.S.D. Europe has some of the most expensive tap water prices worldwide.

The same type of variety risen from EurEau data of Fig. 2.8, representing the annual expenditure per household.

Passing from one ranking to the other, some countries appear in different positions: this could occur because two different types of reference are used (city vs. State) and also for a dissimilarity of the samples associated to each ranking. So that, Italy moves from the last position of Milan in Fig. 2.10 to the second tertile of Fig. 2.11: this means that adopting a higher sample coverage, the country shows a higher water expenditure than its two biggest cities (Rome and Milan).

The Environmental Performance Index (EPI) estimates the impacts caused by states economies on the environment, according to several indicators (Block et al. 2024). According to Fig. 2.12, the top ranked countries are those with the lowest frequency of unsafe sanitation and unsafe drinking water, and they are usually included among those with the highest rates and investments, as shown in the previous figures.

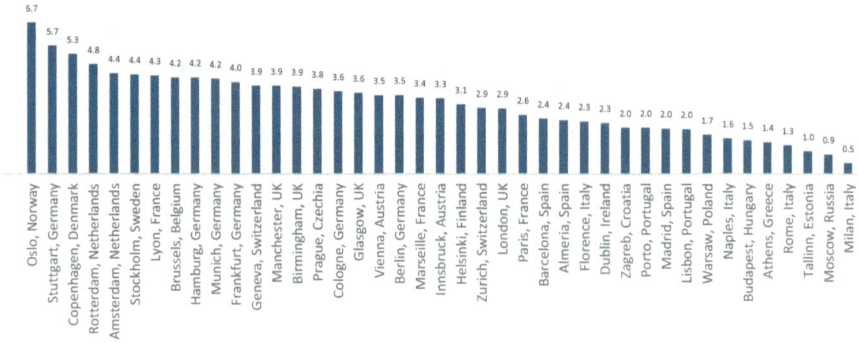

Fig. 2.10 Prices of tap water in selected cities in Europe in 2021 *(in U.S.D. per cubic meter).* (Statista on IBNET)

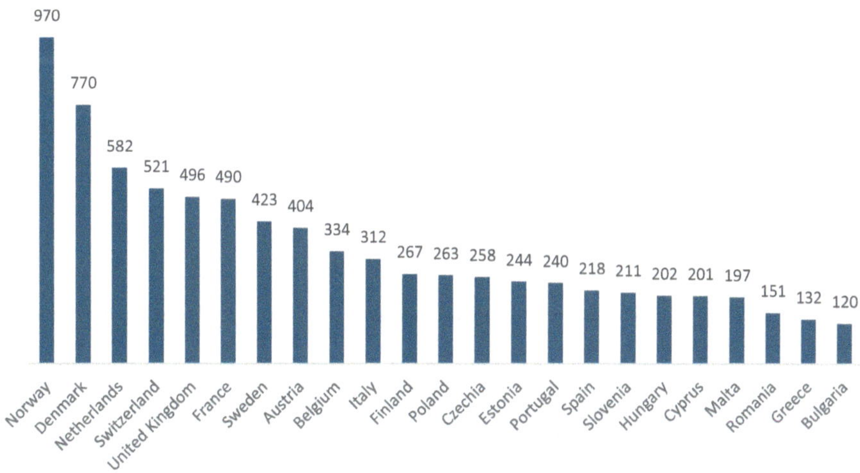

Fig. 2.11 Annual average water bill per household in selected European countries from 2017 to 2019* (in euros). (Statista on EurEau data)

2.2 The Role Played by Water Regulatory Authorities (WRA)

The Water Framework Directive (WFD) 2000/60/EC was established to ensure the efficient use of water resources and the protection of water consumers. One of its key principles is the Full Cost Recovery (FCR), which require to cover all relevant costs associated with water services provisions. EU Communication COM (2000) 477 further clarifies the specific costs that should be covered through tariffs, including operation and maintenance, capital, and environmental and resource costs. The WFD requires the regulation of water prices charged to consumers to ensure cost recovery and, at the same time, to avoid the coverage of inefficiencies of operators.

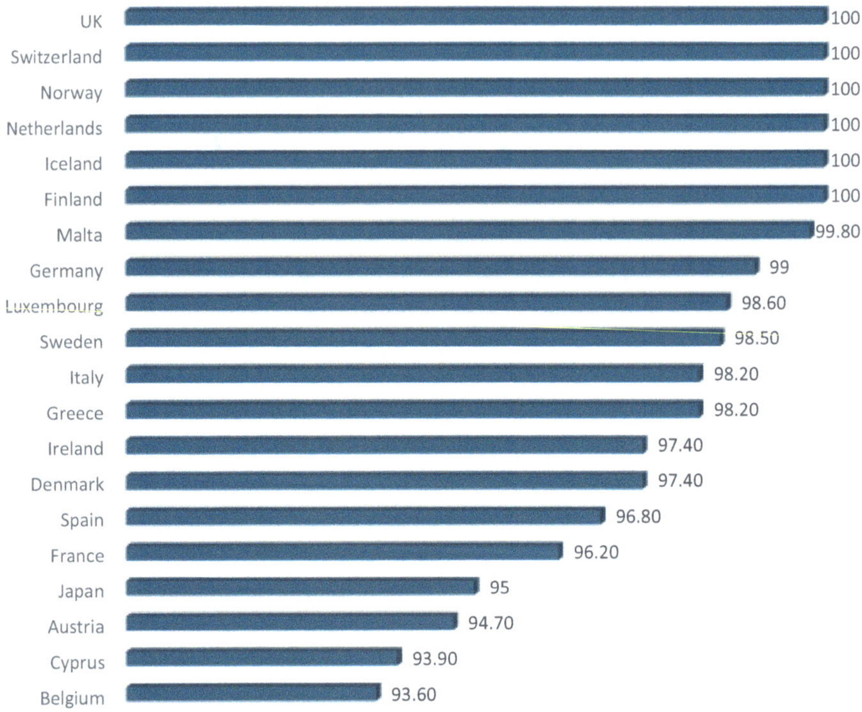

Fig. 2.12 Ranking of countries with the highest sanitation and drinking water score based on the EPI in 2022. (Statista on Eurostat data)

Although the WFD does not provide detailed rules for tariff-setting methodologies, it has prompted many EU member states to assign responsibility to public entities for developing such methodologies. Consequently, independent authorities, water agencies, and ministerial offices across Europe act as economic regulators with varying powers. According to WAREG (2021), most members are entrusted with tariff powers that enable them to define cost components, establish and update tariff methodologies, and approve tariffs. These powers aim to protect consumers and prevent that utilities operating under the conditions of a natural monopoly provide low quality services at a price which doesn't maximize the welfare of the water system. Some water regulators also monitor the technical and contractual quality of service and assign rewards and penalties based on operators' ability to achieve specific targets. Additionally, WRA are involved in setting unbundling rules, monitoring and approving investment plans, granting and revoking licences, implementing measures to ensure affordability, protecting water consumers, enforcing regulations, and addressing operator non-compliance.

The organisational structure of WRA is influenced by the diverse range of tasks they undertake, leading to variations in their independence from national governments, size, and level of specialisation. Some countries, such as Ireland, Italy, and

Portugal, have established authorities with a certain degree of independence from other public entities. These authorities have the autonomy to determine their organisational structure, make hiring decisions, and are financed through fees paid by regulated entities without state subsidies. In these cases, the appointment of board members follows a strict procedure to ensure their independence (OECD 2017). Meanwhile, other countries, such as Hungary, have created agencies that are partially dependent on governmental decisions regarding their organisation and the economic rules they adopt. Finally, in some instances, economic regulation is directly conducted by ministerial offices, maintaining a direct link with the government; examples include Greece (since 2023), Flanders (in Belgium), and Spain. The number of sectors regulated by these entities also varies across countries, with some opting for a multisectoral authority that regulates multiple infrastructural services such as electricity and gas transmission and distribution, district heating, and waste collection and treatment. Others have chosen specialised entities that primarily focus only on water services.

The tariff methodology serves as the primary regulatory tool employed by WRA. The application of the cost recovery principle varies depending on the approach taken by each Member State within the framework provided by the WFD. WAREG (2019) offers a detailed analysis of the tariff methodologies adopted by WRA, highlighting the wide variety of rules in place. The tariff frameworks range from "cost-plus" (7 out of 23 cases) to rate of return (4 out of 23), price cap and revenue cap (8 out of 23), and other approaches (4 out of 23), so grouping different incentive schemes for the European water utilities, with some schemes prioritising the investment maximisation, as the case of the rate of return, while others provide also efficiency targets, as occurring for price and revenue caps.

The rate of return and cost-plus methods are used to determine the allowed revenues for water utilities based on the costs incurred during a specified period, which is usually the most recent financial period for which data are available. Although the regulatory period can be relatively short, sometimes just a year, these methods are designed to ensure that water utilities fully recover their incurred costs. These methodologies do not provide direct incentives for cost optimisation but encourage operators to make investments and get a related level of remuneration.

Conversely, the revenue/price cap method covers a longer regulatory period, typically ranging from three to 6 years, and establishes allowed revenues/prices based on planned costs and business plan analysis. In addition to cost control, regulators often set service quality targets. Under this methodology, the maximum revenue/price is set for the regulatory period, and these variables are adjusted over time to account for inflation (RPI) and promote efficiency growth (X). The inclusion of an efficiency factor (RPI-X) acts as an incentive to reduce costs. The price cap approach includes volume risk, meaning that water utilities bear economic losses from demand shortages unless volume adjustments are implemented.

In addition to the general regulatory framework, the tariff methodologies adopted at country level may include various incentive schemes, some of which may be specifically linked to individual cost items such as energy costs. These schemes aim to promote efficiency in the purchase and consumption processes of water utilities.

2.3 The Public Interventions to Promote a Sustainable Water Sector

The public interventions to boost investments for promoting environmental sustainability and circular economy include: market and governance reforms; economic incentives; strict legal provisions requiring corporate procedures more oriented toward the green economy paradigm.

Market and governance reforms may include several interventions on different segments of the supply chain introducing new market platform and players. These could be based on the involvement of state-owned entities to create value for the material recovered by utilities performing circular based-economy practices: this choice was made in Netherland with the institution of an entity controlled by 14 Dutch water utilities for the disposal of residuals and for their optimum valorisation in the resources market. By contrast, in the UK, the national water regulator Ofwat instituted a new market platform called the "bioresources market", which matches the supply and demand of residuals. This choice counterbalances the strict boundaries adopted by Ofwat to prevent possible subsidization of the circular economy through increased water tariffs, since the regulator aims towards the market development of residuals instead of promoting circular economy through public subsidies. Indeed, the bioresources market website should provide public information from water utilities on the capacity of facilities available to third parties.[5]

The market reforms also include the cap-and-trade mechanism, similar to that developed at EU level for greenhouse gas emissions by energy-intensive factories or the white certificates mechanism for promoting the achievement of energy efficiency targets,[6] which only partially concerns the water sector.

Economic incentives include several measures for allowing utilities to recover resources and costs from investments in circular economy-based activities and to maintain a margin. These measures are both regulatory and non-regulatory:

- public subsidies and grants to develop research projects, such as those issued by the European Commission with LIFE + and Horizon 2020;
- subsidized loans granted by public banks, such as the EIB, for financing "green investments";
- new tariff items introduced by regulators to recover extra costs incurred by utilities to develop circular economy-based activities;
- extra premiums allocated by public authorities for the achievement of environmental goals, as occurs though outcome-based regulation.

Finally, the introduction by **legal provisions** of compulsory behaviour with greater focus on environmental sustainability is also a tool widely adopted for the water sector.

[5] https://www.ofwat.gov.uk/regulated-companies/markets/bioresources-market/bioresources-market-information/.

[6] https://www.gse.it/servizi-per-te/efficienza-energetica/certificati-bianchi.

Type of Interventions		Actors	Strength	Drawbacks
Command and control		Law makers (Parliaments and Governments)	Timeline	Risk of economic losses caused by the behaviour required
Economic incentives	*Regulatory tool*	Independent authorities, ministerial agencies	Timeline; Reduced risk for companies thanks to cost recovery rules	Risk of reducing market engagement in a circular economy and increasing the tariff burden for customers
	Non-regulatory intervention	Public bodies and banks	Lending policies could have a higher economic efficiency than grants	Long-term horizon; Market risk borne by companies in case of borrowing.
Market and governance		Law makers, regulators, utilities	High economic efficiency	Long-term horizon; Market risk borne by companies

Fig. 2.13 Interventions to promote a circular economy (Guerrini and Manca 2020)

For example, the environmental limits set for effluent discharge and sludge disposal (in Switzerland the recovery of phosphorus from the treatment of sewage sludge is required by law, as shown in the European Sustainable Phosphorus Platform[7]) and those set for green procurement by some laws, as occurred in Italy with the reformed code for public procurement.[8]

The three clusters of interventions mentioned above differ in terms of economic sustainability and timeline in the promotion of green actions. Market and governance reforms are based on a long-term horizon and they have a concrete impact on corporate strategy only when the activities promoted are able to generate economic value in terms of circular economy: the slowdown in growth of bioresource market initiatives in UK bears witness to the complexity of creating a new market, counterbalanced by a low public expenditure and greater economic efficiency, once the market will be effectively working. By contrast, the legal imposition of green behaviours acts in the short-term without considering the real economic value for companies of adopting such behaviour. Interventions based on the subtype of economic incentives provided by regulators could balance the timeline of "by-law" actions and ensure at least the coverage of extra costs incurred by utilities, sometimes recognizing tariff increases, although with lower economic efficiency than that associated with a market engagement approach (Fig. 2.13).

2.4 Economic Regulation and Environmental Sustainability

The requirements recently set by European Union, with the recast of the UWWTD, are going to extend the scope of action of water utilities, as they will act as energy prosumers, with the possibility to also become energy retailers in the market, selling

[7] https://www.phosphorusplatform.eu/.

[8] D.lgs. 36/2003.

any surplus of electricity, gas or thermal energy coming out from wastewater processes.[9]

The recognition of the energy potential of the wastewater sector recently made by the new UWWTD is a significant step forward in facilitating a green transition and fostering a more circular economy. As a matter of fact, this new legislation is driving investments towards energy savings, energy production and self-consumption of renewable energy. In addition, the national strategies for energy and climate are also pushing, in some cases, towards climate neutrality in the water and wastewater sector.

Taking stock of these challenging objectives, WRAs have to address the trade-off between climate and energy neutrality on the one hand, and tariff affordability on the other hand: the inclusion of a new range of investments related to energy and climate issues might generate the risk of significant tariff increases which could finally make the water expenditure less affordable and sustainable for households and firms.

Hereafter are described some regulatory insights to accelerate the process of energy transition in the water and wastewater sector, for the achievement of environmental sustainability and affordability goals.

Energy efficiency may be achieved through the reduction of energy consumption of wastewater treatment plants as required by the new UWWTD. However, several policies and regulatory interventions at country level have already anticipated the implementation of this ambitious directive, by supporting and promoting investments linked to energy efficiency. Besides the adoption of several KPIs, some WRAs have adopted in some cases an incentive scheme to valorize the energy saved, giving the possibility to the operator to keep at least a share of these savings, passing on the rest to customers through tariff reduction. Then, with the possibility to include in the Regulatory Asset Base the value of the CAPEX for energy efficiency, a Cost Benefit Analysis could allow to evaluate the presence of a positive variance between the present net value of the extra depreciation, given by the incremental value of an energy efficient technology, and the value of the energy saved and other monetized benefits for the water system. In order to maximize the welfare of the whole sector, regulators should demand a selection of the intervention for energy efficiency, to avoid the support of inefficient technologies: as a consequence, before assigning a reward or a profit sharing for energy savings, regulators should properly evaluate in detail the projects which are going to be implemented. This would imply an extra regulatory burden, which needs to be shared with local water

[9] "The urban wastewater treatment sector could play an important role in significantly reducing greenhouse gas emissions and helping the EU achieve its climate neutrality objective. The new rules introduce an **energy neutrality target**, meaning that **by 2045** urban wastewater treatment plants treating a load of 10,000 population equivalents and above will have to use energy from renewable sources generated by the respective plants." https://www.consilium.europa.eu/en/press/press-releases/2024/11/05/urban-wastewater-council-adopts-new-rules-for-more-efficient-treatment/.

authorities, where a multilevel governance is in force, or with extra resources assigned to the WRA.

Beside this type of incentive scheme, a cap-and-trade mechanism could be implemented at country or even at European level, to ensure the implementation of energy savings interventions only where it is efficient. Utilities would be obliged to purchase energy credits from those who exceeded the target assigned. This approach is quite similar to the Emission Trading System scheme and to the "white certificate" model in force in Italy for energy efficiency, which, unfortunately, is not mandatory for water utilities.

Regarding energy neutrality, regulators could set up some economic levers, besides KPIs, to incentivize the implementation of dedicated investments, while also fostering cost efficiency. Whereas the UWWTD requires energy neutrality regardless of its costs, regulators could support the achievement of these targets while addressing tariff affordability challenges. The renewable power plant might not always achieve a levelized cost of energy (LCOE) lower than the average market prices of energy. In case of rising of a positive difference between those terms, the purchase of green energy would be a more efficient solution than the internal production. Following this idea, the amended and promulgated version of the UWWTD allows the purchase of non-fossil energy up to 35%.

The incentives which could be put in place by WRA range from the inclusion of the cost of energy plants in the Regulatory Asset Base to the valorization of self-produced and consumed energy expressed in €/kWh. The first option can be a powerful incentive, avoiding any loss for the utility, passing on to customers the risk of positive differences between the LCOE and market prices. Thus, this solution may not be the most efficient one and could be adopted for a transitory period to trigger the energy transition of the sector. Once the adoption of renewable power plants would be sufficiently deployed, a regulator could then move on to a more efficient incentive option, which bases the recovery of the total cost of ownership on the repayment of the kWh self-consumed according to a allowed price linked to the energy market. This incentive scheme could be designed adding a sharing mechanism of any variance between the allowed price and plant's LCOE with customers.

Moreover, when utilities sell the surplus of energy produced to third parties, regulators could consider specific methods to treat those extra revenues. In a first stage, the implementation of a "single till" method could assure the compliance with the full cost recovery, and assure a powerful incentive for renewables plants installation. Then, the regulator could ask to the utility to share with customers at least a part of the revenues coming from the sales of energy, adopting an "hybrid till" method.[10] Finally, when the practice of renewable energy production is widespread, a "dual till" method would ensure efficiency and tariff affordability for

---

[10] "The Italian Authority recently identified in the "other water-related activities" category …a list of innovative activities related to energy and environmental sustainability…including, among others, raw material recovery from water and wastewater treatments, providing for the in-tariff reimbursement of a higher margin achieved by the operator (75%), increasing the profit sharing of these activities" (Guerrini and Manca 2020).

water customers: under this approach, the cost of water/wastewater and energy services are fully unbundled, so that the energy service must be self-sustained from an economic perspective, without any possible extra charge being passed on to water customers.

Finally, climate neutrality represents the compensation of GHG emissions generated by the sector. Such compensation could include interventions not strictly related to the water and wastewater sector, to produce positive environmental externalities, such as nature-based solution, reforestation, reuse of treated wastewater for urban irrigation and street cleaning or use of electric vehicle fleet. Despite their positive externalities, Usually these interventions do not directly generate revenues nor savings for water and wastewater utilities, as this is the case with energy efficiency and energy production choices, and they must be fully recovered through water tariff or environmental taxes. The possibility to include the costs of these types of investment into the environmental costs introduced by the WFD could be a key step towards charging the full cost of climate neutrality actions to water consumers. Through this arrangement that accounts for positive environmental externalities, all water users will thus pay for the benefits deriving from less hydrogeological instability, better quality of air and soil, improved water quality and replenishing of water bodies.

2.5 The Main Performance Drivers Affecting the Water Sector

According to the scientific literature and to practical experience, the water sector is affected by some key factors, to which the performance of utilities is related: firm's size and the size of the area served, population density and other water governance issues could drive the performance of operators and the cost and quality of services experienced by customers. The next subparagraphs illustrate the main literature evidences on these issues.

2.5.1 The Economies of Scale

Given a production function, economies of scale arise when a unit increase in output results from a less than proportional increase in input.

Most studies confirmed the presence of economies of scale in the water industry (e.g., Shih et al. 2006; Guerrini et al. 2011; Carvalho and Marques 2014), but several others found diseconomies of scale in various countries (e.g., Bhattacharyya et al. 1995; Saal and Parker 2000; Saal et al. 2007;). Several studies on economies of scale show that only small- and medium-sized utilities can improve efficiency through expansion, and that larger utilities do not always benefit from expansion, and sometimes, even suffer diseconomies of scale (Torres and Morrison-Paul 2006; Filippini et al. 2008; De Witte and Marques 2011).

Researchers have not agreed on an "optimal scale" for water utilities (Guerrini et al. 2013), which appears to vary considerably among countries (Saal et al. 2013).

Fabbri and Fraquelli (2000) finds large economies of scale for utilities that deliver a minimum of around 350,000 m^3 of water and diseconomies at the maximum point of 394 mn. m^3. Focusing on multi-utilities, Fraquelli et al. (2004) finds economies of scale only for utilities with output levels lower than those at the median size; those bigger than the median experienced neither economies nor diseconomies of scale. Guerrini et al. (2018) demonstrates that economies of scale affect the water sector, benefiting utilities serving less than 50,000 customers as well as those serving between 50,000 and 150,000 customers, with a decreasing elasticity.

2.5.2 *The Economies of Density*

The literature has also addressed economies of density (Caves et al. 1984). In the water industry, economies of density exist in the presence of greater population density or of an increase in the water provided per kilometer of mains, because the cost of the infrastructure required to provide the service is lowered. Fabbri and Fraquelli (2000) finds that greater customer density leads to cost savings: an effect confirmed by Antonioli and Filippini (2001) and Guerrini et al. (2013).

Guerrini et al. (2018) shows that economies of density affect utilities serving between 30 and 58 customers per kilometer of network, demonstrating that utilities operating in densely populated areas such as cities achieve the lowest cost for delivering a cubic meter of water. This cost advantage depends on the ability to deliver a cubic meter of water with fewer resources (i.e., mains and electricity), and to limit water losses per customer. These findings are in line with the previous literature that examined this issue for various countries (Fabbri and Fraquelli 2000; Antonioli and Filippini 2001; Garcia and Thomas 2001; Tupper and Resende 2004; García-Sánchez 2006; Torres and Morrison-Paul 2006; Filippini et al. 2008; Nauges and van den Berg 2008). Some diseconomies could appear when water leakages are measured, since leakages per km grow up with the growing number of connections handled by a water utility (Van den Berg 2014).

2.5.3 *Water Governance Issues*

Starting from '90 s, many countries implemented structural reforms of public sectors, through privatization of national monopolies; unbundling of separate activities of the supply chain (management of infrastructure and retail); institution of an independent authority.

A different institutional framework could interfere with the performance achieved by public utilities. Cambini and Spiegel (2016) shows that regulated utilities realize more investments and are more oriented to borrow. The "dash of debt" of regulated entities is widespread across countries and sectors (electricity, gas, water). It rises from the strategy of utilities to reduce the risk of opportunistic behaviour or regulators which could initially set quite advantageous conditions to incentivize investments of private utilities and then, once investments have been realized, could made some tariff cuts. A management strategy to avoid the negative consequences for utilities deriving from this circumstance is based on the increase of financial leverage, so that the likelihood of a tariff cut will be reduced when it implies a risk of financial distress (Cambini and Rondi 2006).

Another driver of financial leverage is based on the method applied by WRA to set the gearing for WACC estimation purposes (Guerrini 2024). A WRA can align the value of the notional gearing adopted in the WACC with the average gearing value of the sector, or alternatively with that of the companies in the first percentile, with lower gearing, or in any case with a better credit rating. This second approach is applied, for example, by the water regulator OFWAT, which defines bankability requirements within the Price Reviews, setting the range with which the rating of an ideal company[11] must vary and associating a certain debt value with it, which in turn is used by the rating agencies to estimate the scoring to be assigned to companies in the industry.

The demands of regulated companies vary based on WRA's approach. When WRA emphasizes notional gearing for cost recovery, companies arbitrate between the notional returns and the actual financial costs. Conversely, if WRA links debt rates to a notional rating, companies aim to maintain favorable bankability indexes. Reducing notional gearing could hinder credit access for operators with varying values.

A hybrid approach could mitigate this trade-off by using the sector's average indebtedness rate to define the regulatory WACC, while also identifying an optimal indebtedness level for the notional rating.

While the WRA can affects multiple aspects of corporate management - including operational efficiency, investment planning, service quality, procedures for measuring user consumption, innovation and environmental sustainability—the decisions on corporate finance have instead always been considered under the full discretion of companies, leaving room for the intervention of the "dash for debt" effect. Then, the regulatory decision to avoid any intervention in the financial structure gives rise to the risk of having a sub-optimal management of financing sources, oriented towards short-term profit maximisation, in order to ensure high payouts to shareholders within the board's mandate. While in Italy the debt values of regulated companies in the water sector are rather low, as shown in Fig. 2.14, in other countries, utilities have shown a greater propensity to build financial structures with a predominance of debt (Fig. 2.15).

[11] In PR24, the company's 'notional' rating was set at BBB+/Baa1.

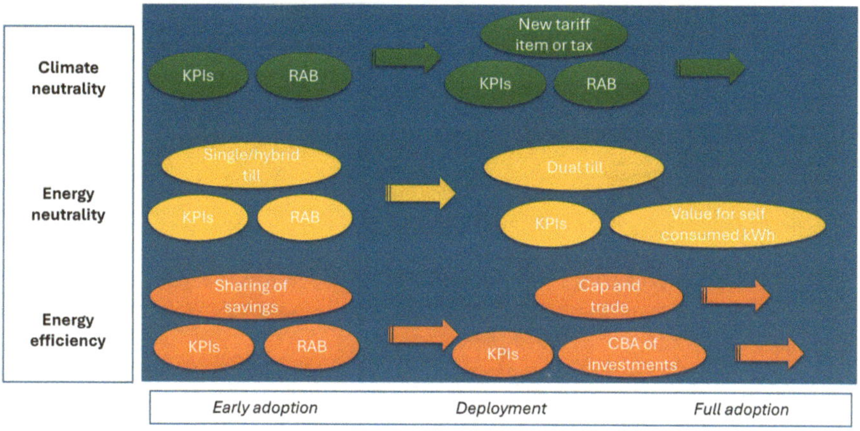

Fig. 2.14 Regulatory tools to support the energy efficiency and climate neutrality transition

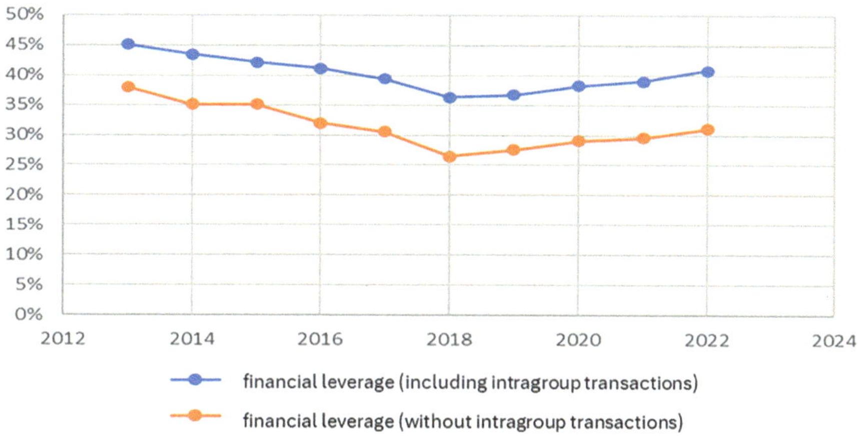

Fig. 2.15 The debt index in Italian utilities (*Blue Book* 2024, *Utilitatis*)

Figure 2.16, published by The Guardian in an article on the Thames Water case,[12] highlights the orientation of UK fully private utilities to increase the weight of debt on the overall structure of funding sources, reaching gearing levels beyond the notional values defined by OFWAT.[13] Thames Water's gearing, for example, was around 20 percentage points higher than the regulatory gearing. In general, the risk of default affecting some water utilities in the UK would seem to be attributable to

[12] *In charts: how privatisation drained Thames Water's coffers*, The Guardian, 30th of June, 2023, https://www.theguardian.com/business/2023/jun/30/in-charts-how-privatisation-drained-thames-waters-coffers.

[13] The figure shows the gearing value prior to PR24, where a further 5% reduction was applied by OFWAT.

Fig. 2.16 The gearing level of UK water utilities (The Guardian)

the excessive use of leverage which, in combination with the inflationary wave of 2022, generated an increase in debt servicing cash flow. Approximately half of Thames Water's more than £ 15 bn of debt is indexed to inflation and has therefore led to serious repayment difficulties for the operator, with a connected trouble in meeting the financial requirements associated with the realisation of investments and, more generally, the achievement of the quality standards assigned by the WRA. Taking into account that the company operates in a macro-area covering 16 mn customers, which includes the city of London and the Thames basin, its crisis situation is a serious problem for the country, with a possible intervention by the central government, also in the light of the recent request of the company for a tariff increase of more than +50%, which has not been accepted by the regulator.[14]

Literature revealed that most regulated firms pay out a higher portion of their earnings than firms operating in non-regulated industries and that high corporate payouts are the outcome of higher transparency and shareholder protection induced by a clear and predictable tariff-setting method. Therefore, regulation may influence dividends paid by regulated firms. Then, the ownership structure affects the rate of payout, with the highest rate achieved by private owned utilities (Romano and Guerrini, 2019).

The lack of effective regulatory action concerning the financial structure of private owned utilities can therefore lead to several drawbacks that can affect the soundness of an entire industry, with repercussions on the quality of service and the reputation of the companies involved. The WRA must therefore intervene to prevent

[14] *Thames Water's biggest shareholder writes off investment*, Financial Times, 17th of May, 2024, https://www.ft.com/content/aa0a8179-27ce-42aa-9687-b4a61360c907.

an indiscriminate increase in debt, coupled with generous dividend payouts, from creating liquidity crises, which could jeopardise the capability to realise the investment planned.

References

Antonioli B, Filippini M (2001) The use of a variable cost function in the regulation of the Italian water industry. Util Pol 10(3–4):181–187

Bhattacharyya A, Harris T, Narayanan R, Raffiee K (1995) Technical inefficiency of rural water utilities. J Agric Resour Econom 20:373–e391

Block S, Emerson JW, Esty DC, de Sherbinin A, Wendling ZA et al (2024) 2024 environmental performance index. Yale Center for Environmental Law & Policy, New Haven, CT. epi.yale.edu

Cambini C, Rondi L (2006) Struttura finanziaria e investimenti in un'impresa regolata. Polit Econom 3(2006):317–360

Cambini C, Spiegel Y (2016) Investment and capital structure in a partially privatized utility. J Econom Manag Strat 25(2):487–515

Carvalho P, Marques R (2014) Computing economies of vertical integration, economies of scope and economies of scale using partial frontier nonparametric methods. Eur J Oper Res 234(1):292–307

Caves DW, Christensen LR, Tretheway JA (1984) Economies of density versus economies of scale: why trunk and local service airline cost differ. Rand J Econ 15(4):471–489

De Witte K, Marques RC (2011) Big and beautiful? On non-parametrically measuring scale economies in non-convex technologies. J Prod Anal 35(3):213–226

EurEau (2021) Europe's Water in Figures. www.eureau.org

Fabbri P, Fraquelli G (2000) Costs and structure of technology in the Italian water industry. Empirica 27(1):65–82

Filippini M, Hrovatin N, Zori J (2008) Cost efficiency of Slovenian water distribution utilities: an application of stochastic frontier methods. J Prod Anal 29(2):169–182

Fraquelli G, Piacenza M, Vannoni D (2004) Scope and scale economies in multi-utilities: Evidence from gas, water and electricity combinations. Appl Econ 36(18):2045–2057

Garcia S, Thomas A (2001) The structure of municipal water supply costs: Application to a panel French local communities. J Prod Anal 16(1):5–31

García-Sánchez IM (2006) Efficiency measurement in Spanish local government: The case of municipal water services. Rev Pol Res 23(2):355–371

Guerrini A (2024) La struttura finanziaria aziendale. Il "dark side" della regolazione economica. Azienditalia—Mensile di organizzazione, bilancio, gestione e controllo degli enti locali, pp 8–9

Guerrini A, Manca J (2020) Regulatory interventions to sustain circular economy in the water sector. Insights from the Italian regulatory authority (ARERA). H2 Open J 3(1):499–518

Guerrini A, Romano G, Campedelli B (2011) Factors affecting the performance of water utility companies. Int J Public Sect Manag 24(6):543–566

Guerrini A, Romano G, Campedelli B (2013) Economies of scale, scope, and density in the Italian water sector: a two-stage data envelopment analysis approach. Water Resour Manag 27(13):4559–4578

Guerrini A, Romano G, Leardini C (2018) Economies of scale and density in the Italian water industry: a stochastic frontier approach. Util Pol 04(52):103

Nauges C, van den Berg C (2008) Economies of density, scale and scope in the water supply and sewerage sector: a study of four developing and transition economies. J Regul Econ 34(2):144–163

OECD (2017) The governance of regulators creating a culture of independence practical guidance against undue influence. OECD. www.oecd.org

OECD (2020) Financing water supply, sanitation and flood protection challenges in Eu member states and policy options. OECD. www.oecd.org

Saal D, Parker D (2000) The impact of privatization and regulation on the water and sewerage industry in England and Wales: a translog cost function model. Manag Decis Econom 21:253–e268

Saal D, Parker D, Weyman-Jones T (2007) Determining the contribution of technical, efficiency and scale change to productivity growth in the privatized English and welsh water and sewerage industry: 1985-2000. J Prod Anal 28:127–139

Saal D, Arocena P, Maziotis A, Triebs T (2013) Scale and scope economies and the efficient vertical and horizontal configuration of the water industry: a survey of the literature. Rev Netw Econ 12(1):93–129

Shih J, Harrington W, Pizer W, Gillington K (2006) Economies of scale in community water systems. J Am Water Works Assoc 98:100–e108

Torres M, Morrison-Paul CJ (2006) Driving forces for consolidation or fragmentation of the US water utility industry: a cost function approach with endogenous output. J Urban Econ 59(1):104–120

Tupper H, Resende M (2004) Efficiency and regulatory issues in the Brazilian water and sewerage sector: An empirical study. Util Pol 12(1):29–40

Utilitatis (2024) Blue Book. www.utilitatis.org

Van den Berg C (2014) The drivers of non-revenue water how effective are non-revenue water reduction programs? World Bank Group Policy Research Working Paper, p 6997

WAREG (2019) Tariff Regulatory Frameworks in WAREG Member Countries. www.wareg.org

WAREG (2021) Water Regulatory Governance Across Europe. www.wareg.org

Chapter 3
Materials and Method

This report has multiple aims.

It provides an overview of the performance of the private owned utilities in the European water sector, including mixed owned companies, joining private and public shareholders in the same entity through the arrangement of Public-Private Agreements; and wholly private owned companies.

The analysis of financial highlights allows to get some contributes on the attractiveness of the European water sector for private investors, showing the trend of rates of return and their main drivers (profitability, turnover, financial leverage).

Then, the measures observed are used also to identify the effects of some institutional variables, such as the country features, the water governance asset, including regulatory affairs by clustering countries with independent and national regulators, and those operating without this model.

Finally, the effect of the scope of action is observed, clustering companies providing only water services and those operating with an integrated business model (water and wastewater management).

The selected panel includes 32 entities located in 10 European countries, and serving approximately 50.5 mn of inhabitants, round the 10% of the sum of the European and England and Wales population: considering that a large part of population in Europe is served by municipalities or state owned companies, this rate could be quite significant, giving a good representative power to the panel.

The list of companies has been identified through a multiple-selection process, which began with a web site overview per country, looking for water entities controlled by private investors. Specialized report as well as information provided by water associations website allowed to identify a first list of companies per country, which was then double-checked with the support of ORBIS (a dataset which provides financial information for legal entities with a worldwide perspective) for monitoring the company's going concern during the observed period (2028–2023) and removing some outliers with anomalous trends (Fig. 3.1).

© The Author(s), under exclusive license to Springer Nature Switzerland AG 2026

A. Guerrini, *Private Investments in the European Water Sector*, SpringerBriefs in Water Science and Technology, https://doi.org/10.1007/978-3-032-24735-3_3

Fig. 3.1 The features of the panel

Country	Nr. of companies	Inhabitants
Bulgaria	1	1,400,000
France	5	949,595
Germany	3	1,014,264
Greece	1	1,200,000
Hungary	1	200,000
Italy	5	6,167,415
Portugal	5	720,113
Spain	6	11,430,695
United kingdom	4	27,000,000
Czechia	1	380,000

The country features are described in the Annex II, while the list of companies selected is showed in Table 1 of the Annex III.

The regulatory profile of each country is quickly summarized by Fig. 3.2, which represents in blue those countries with a national WRA, and in blue those countries with a regional or municipal regulator, or without a WRA.

Considering the segment, while 26 out of 32 utilities operate in the integrated water service, the rest are only engaged in the water segment (Fig. 3.3).

Once the panel has been designed, a list of indicators was gathered for the indicated six years period, using ORBIS.

The performance has been enquired applying the Return on Equity (ROE) pyramid.

A regulated company has multiple levers to enhance its return, many of which derive from the incentive schemes defined by the WRA and the managerial ability to act on specific performance drivers, such as: increasing operational efficiency and achievement of service quality targets to which reward schemes are linked (acting on Return on Investments (ROI)—Return on Asset (ROA)); fiscal strategies; and, finally, financial management (acting directly on ROE),

The last driver intervenes on returns through the leverage, with the increase (or decrease) in net return on equity due to debt increases (decreases), in the presence of a positive spread between ROI and cost of debt (Cd). At the same time, financial management influences the cost structure through the cost of sources and its gap with the notional financial cost, as defined by the WRA.

The increase in return on equity is represented following the Miller and Modigliani (1963) following formula[1]:

[1]The following formula, presented in the handbook on financial statement analysis, was developed by Modigliani and Miller and corresponds to the second proposition of the theorem of the same name. See the following contribution for a correction of the second proposition to take account of the fiscal impact: Miller, M.; Modigliani, F., 1963. *Corporate Income Taxes and the Cost of Capital: A Correction*, in *American Economic Review*, vol. 3, no. 53, p. 433–443.

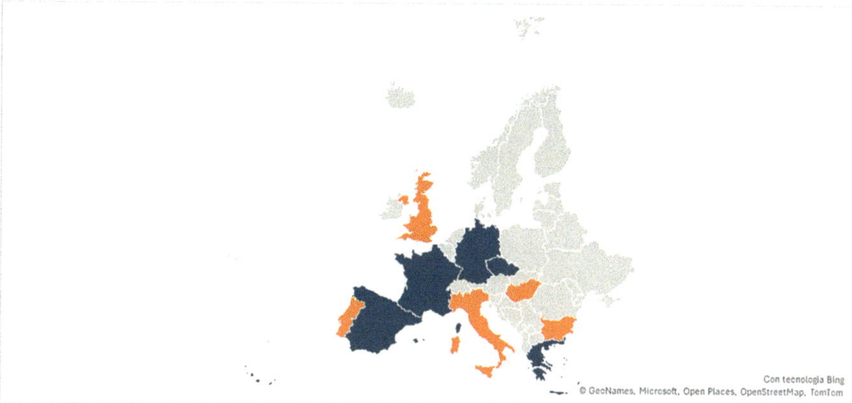

Fig. 3.2 The regulatory profile of the countries included in the panel

Fig. 3.3 List of investors associated to the number of inhabitants provided

Investors	Inhabitants
Stocks exchange	22,500,000
Veolia	10,363,417
ACEA	5,900,171
Citigroup	5,700,000
Eugenio Calabuig Gimeno	3,000,000
Suez	913,144
E,ON	600,000
ANTIN INFRASTRUCTURES SERVICES	435,000
Saur	285,113
ACEA-SUEZ	267,244
Acciona	220,000
Vinci	186,729
Eurawasser	91,264

$$ROE = \left[ROI + \left(ROI - Cd\% \right) x \frac{D}{E} \right] x \left(1 - T \right) \tag{3.1}$$

In which:

ROE = Net profit/Shareholders' equity.
ROI = EBIT/Total assets.[2]
cd% = the cost of debt.

[2] In this text the calculation of the ROI is simplified, assuming that all firm's assets concern the operating activities. With this assumption ROI is equal to ROA.

T = tax rate,

The formula can also be represented as follows by inserting the variable 'g' for the gearing:

$$ROE = \left[ROI + (ROI - Cd\%)x\frac{g}{1-g}\right]x(1-t) \quad (3.2)$$

To summarise, the rate of return on equity increases as the debt ratio increases, if there is a positive differential between ROI and Cd: in the presence of a positive variance between the return on invested capital and the cost of debt, it is economically advantageous to increase the share of debt; at the same time, the reduction of the weight of equity in total sources leads to an increase in ROE.

The ROI/ROA drivers are:

- The profitability, given by the ratios of different operating gross margins (EBITDA, EBIT) and net profit to the total revenues or the Value of Production (VoP) measured in this report;
- The turnover ratio, given by the ratio of VoP to the total asset.

The index of profitability measures the margin yield by the firm for each monetary unit of sales or VoP. The rate of EBIT margin represents the economic amount available to remunerate the debt and the equity and to pay taxes. In a regulatory context a large rate shows the firm's ability to act on operational efficiency, in order to keep an higher margin for remunerating the financial structure; it could also derive from an overcompensation of efficient cost by the WRA through the estimation of allowed revenues.

The turnover represents a proxy of the firms' financial cash flow obtained from a certain amount of capital invested. For a regulated entity, a high turnover could represent a high quote of operating expenditure associated with a low propensity to investment realization; in case of some serious flaws in economic regulation, it could also come from an overestimation of allowed revenue by the WRA, comparatively to the activity effectively realized by the utility.

Reference

Miller M, Modigliani F (1963) Corporate income taxes and the cost of capital: a correction. A Econ Rev 3(53):433–443

Chapter 4
Main Results of the Study

4.1 Revenues Per Inhabitant

This section illustrates the data collected for the selected panel, following to the methods described. Further comments and practical implications of the results are presented in the fifth chapter of the report.

The water price is one of the most widely used indexes for providing a first overview of the sector, especially when different countries are compared. Moving from a gap with the highest standards of quality, in many countries the water price is interpreted as a synonymous of the capability to collect funds for investments, so opening the route toward quality improvement. Then, if compared with the country GDP per capita, it also provides information on affordability.

The index is calculated starting from the tariff structure, alternatively articulated in flat price or increasing block prices according to the growth of volume of water consumed. A standard consumption is usually identified for estimating a water expenditure per household.

For the purposes of this report a proxy of water price was measured: given the difficulty to collect information on the tariff structure in each case monitored through the panel, the VoP per inhabitant has been calculated: despite its evident drawbacks—inclusion of capitalized costs and variations in stock of raw materials—it provides a first glance at water expenditure for a customer.

The ranking shown in Fig. 4.1 is actually quite consistent with other analysis, as the one provided in Fig. 2.11. In terms of values, the data should be multiplied for the number of components of a family unit to get the correspondent water expenditure: considering a household of three units, United Kingdom and Germany reach approximately 900 €, Greece is at the bottom with 186 €, and Portugal and France has the median value of approximately 600 € (Fig. 4.1).

© The Author(s), under exclusive license to Springer Nature Switzerland AG 2026

A. Guerrini, *Private Investments in the European Water Sector*, SpringerBriefs in Water Science and Technology, https://doi.org/10.1007/978-3-032-24735-3_4

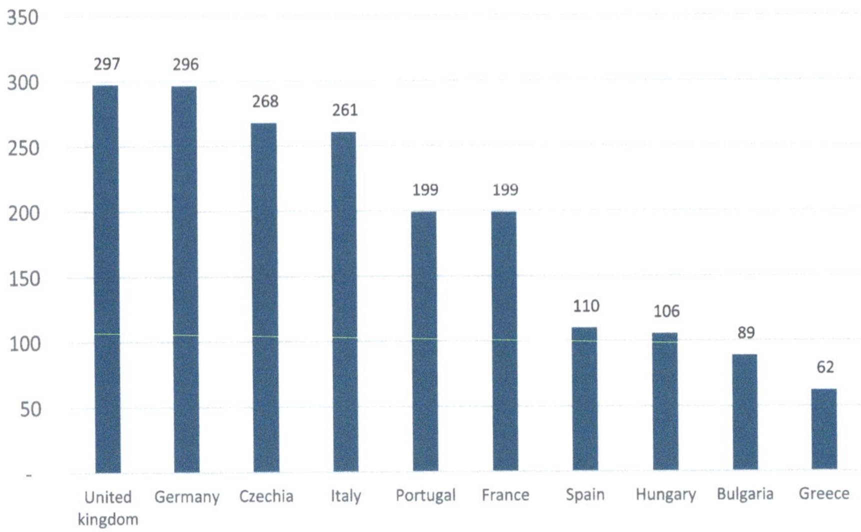

Fig. 4.1 Water expenditure per year per inhabitant per country (€/year/person)

Where a national WRA has been set the water expenditure is higher than other cases: 230 € vs. 180 € of water expenditure per year per person is recorded respectively, with a variance of 50 €.

Looking to the water value chain, the integrated water service demands approximately 100 € more than only water service (224 € vs. 123 €) (Fig. 4.2).

4.2 Investments

The investments trend is enquired using two items:

- the Total Asset (TA)
- the variance of TA per year without the effect of amortization and depreciation (INV).

The first item is available in the firm's financial statements; the second item has to be calculated from TAs of two consecutive years, adding the depreciation of the more recent year: the TA represents the net cumulated book value of investments; the variance of TA shows the level of investment realized in a given year. For the purposes of this report, both indexes have been related to the number of inhabitants to better show the propensity of the operators to realize a significant amount of infrastructures. Some differences among companies could rise also for different accounting rules and license schemes.

The panel shows a wide variance among countries, with the United Kingdom records of 2459 of TApc in 2023, demonstrating a long period of massive investments in the sector. In terms of investments per year, several countries show a

Investors	Water expenditure (€/year/person)
Eurawasser	733
Acea	264
Citigroup	251
Listed Company	250
Acea-Suez	249
Suez	244
Saur	210
Antin Infrastructures Services	192
Eugenio Calabuig Gimeno	151
Vinci	125
Veolia	118
Acciona	109
E.On	94

Fig. 4.2 Water expenditure per year per inhabitant per investor (€/year/person)

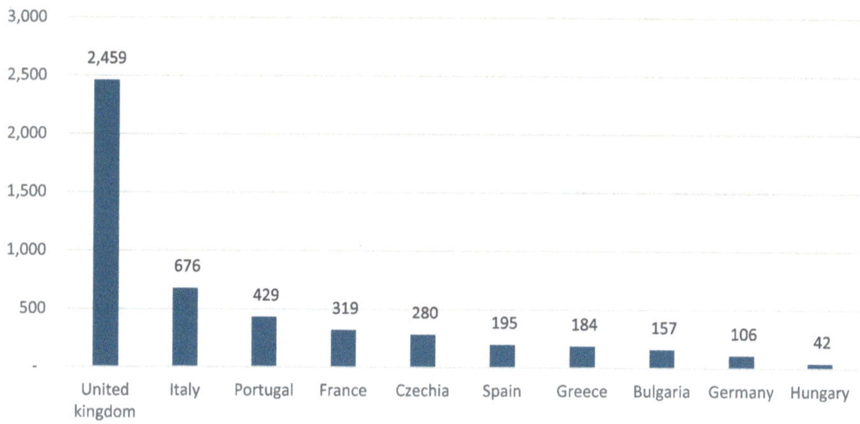

Fig. 4.3 TApc per country (€/2023/person)

positive trend during the period 2018–2023, with United Kingdom and Italy having the highest records of approximately 70 €/inhabitant.

Investment trends seem to be affected not only by the firm's propensity to realize and renovate the water and wastewater infrastructural network, but also by the institutional asset, as represented by the regulatory model chosen. Those countries with a national WRA record an investment trend between 40 and 50 €/inhabitant, while the other countries are placed around 10 €/inhabitant (Figs. 4.3, 4.4 and 4.5).

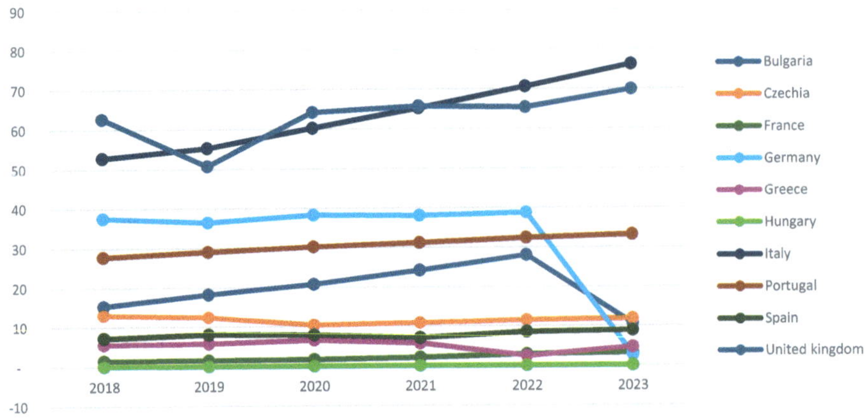

Fig. 4.4 INVpc per year per country(€/year/person)

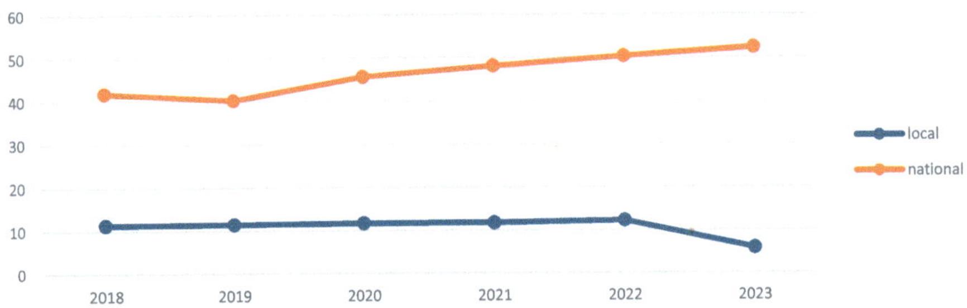

Fig. 4.5 Investments per inhabitant per year per regulatory model (€/year/person)

4.3 Financial Leveraging

The panel has a different composition also in terms of degree of resort to debt for financing the investment needs. United Kingdom has the highest weight of debt (approximately 5 times the equity), followed by Portugal (3 times) and Italy (1 times). Whether a too high value could bring to a firm financial collapse, from the other side, a very low resort to debt could be due to a lack of efficiency or to a low financial need due a poor volume of investment to realize, as it seems to occur for Greece, Bulgaria and Hungary (Fig. 4.6).

The different regulatory model could explain the variety of gearing values, for the reasons described in Chap. 2: the countries of the panel with a national WRA show a gearing value of 244, while the others have an average value of 55, with a debt reaching half of the equity value.

From an investors perspective, the highest value is achieved by Citigroup, with a debt 8 times greater than equity, overpassing the average of the listed companies operating in the same country (UK) (Fig. 4.7).

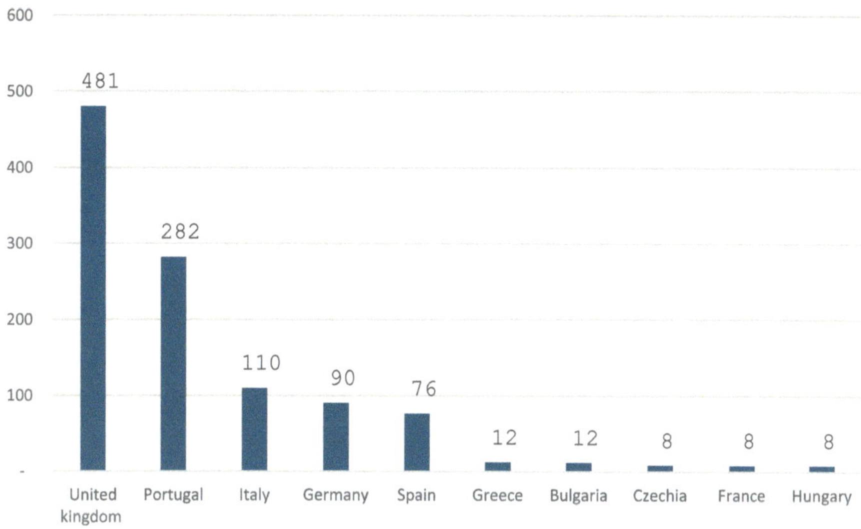

Fig. 4.6 Net gearing per country (%)

Fig. 4.7 Net gearing per investor (%)

Investor	Gearing
Citigroup	805
Antin Infrastructures Services	379
Listed Company	282
Eugenio Calabuig Gimeno	175
Saur	138
Acea	127
Eurawasser	104
Veolia	48
E.On	45
Acea-Suez	42
Acciona	42
Suez	11
Vinci	n.a.

Considering a dynamic perspective, the gearing diminished during the observed period in both type of regulatory context, also in correspondence with the increasing of inflation rate which from one side affected the policies of central banks in rising

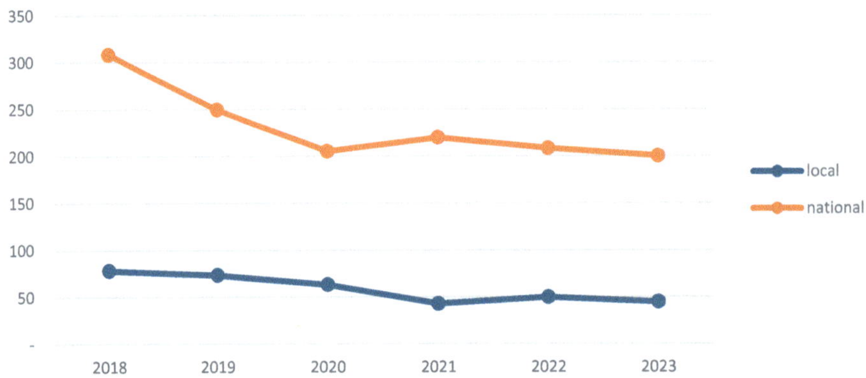

Fig. 4.8 Gearing trend during the observed period

the interest rate, and from the other side it directly determine the growth of the cost of debt for index linked bonds (Fig. 4.8).

4.4 Rate of Returns

This section illustrates the core data for investors, focused on three types of rates of return. While ROA represents the operating return on total asset, ROCE moves from the previous ratio netting of effect of current liabilities; finally, ROE is the well-known rate or return on equity which provides the actual remuneration of shareholders. The next figures provide average values for the period 2018–2023 for countries and types of regulatory model respectively.

ROCE is usually greater than ROA, for the positive effect exerted by current liabilities which cut the denominator; similarly, ROE overcomes the level of ROCE, for the positive effects exerted by the financial policy, with the exceptions of Spain and United Kingdom, where perhaps a spread between the actual cost of debt and ROA could have damaged the ratio.

The variances between "national" and "local" regulatory models are really significant: for firms regulated by a WRA the ROE is approximately 6 times that of firms operating without a WRA or only with locally based regulator. In this second case, the low level of indebtedness determines sometimes a ROE lower than ROCE (Figs. 4.9, 4.10 and 4.11).

During the observed period the spread between both regulatory models remained positive, with the exception of 2019, which represents the worst year for the cluster of companies operating under a "national" WRA; while in 2021 the other cluster recorded a negative performance: in terms of ROA, in 2020 and 2022 this cluster

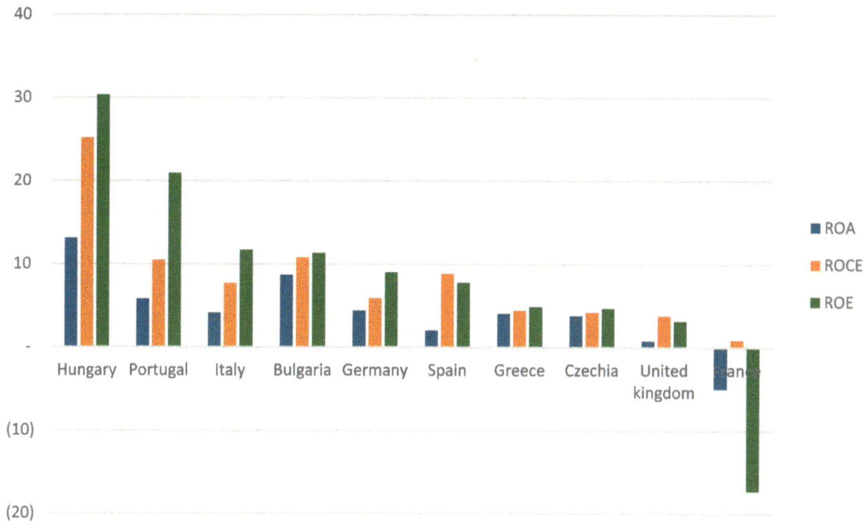

Fig. 4.9 Rates of return per country

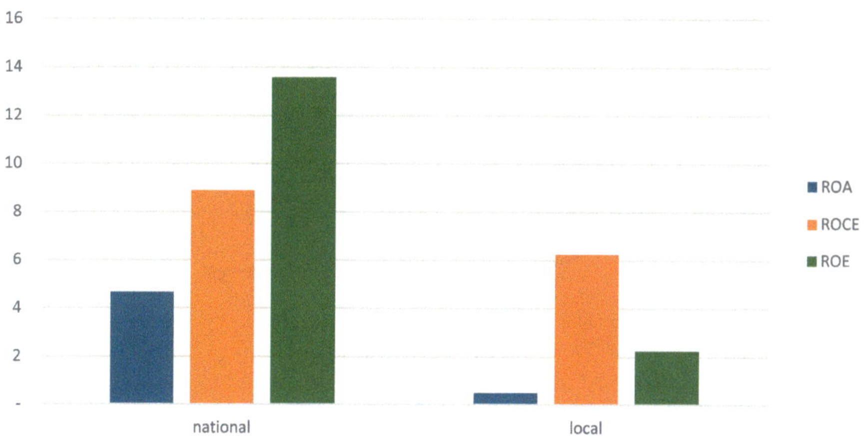

Fig. 4.10 Rates of return per regulatory model

recorded a negative performance; then, the high variance of ratios demonstrates that the lack of regulation could create higher instability (Figs. 4.12 and 4.13).

Investor	ROA	ROCE	ROE
Antin Infrastructures Services	4.70	9.24	21.07
Saur	7.41	12.33	20.72
Eugenio Calabuig Gimeno	5.71	9.30	20.57
Acciona	3.04	36.28	18.57
Acea	4.20	8.00	12.67
Veolia	0.99	7.42	8.03
Acea-Suez	3.99	6.69	7.63
E.On	3.93	5.95	6.11
Eurawasser	1.90	3.08	4.18
Listed Company	1.72	4.04	4.13
Citigroup	0.14	3.20	0.81
Vinci	-2.00		
Suez	1.37	2.00	-15.20

Fig. 4.11 Rates of return per investor

Fig. 4.12 ROE trend during the observed period per regulatory model

Fig. 4.13 ROA trend during the observed period per regulatory model

4.5 Profitability

Following the ROE pyramid, the return on operating investments is driven, from one side, by profitability, measured in terms of differential between revenues and costs.

The spread between EBITDA margin and EBIT margin is directly proportional to the volume of cumulated assets to be depreciated. The first three countries ranked in Fig. 4.14 are those with the highest amount of TA realized per inhabitant. The spread between EBIT margin and Net profit margin is driven by the firm's exposure with banks, measured by the net gearing, and by the cost of debt (Figs. 4.15 and 4.16).

Grouping countries according to their regulatory model, a positive spread on profitability index features the WRA comparatively with local based model; this gap is gradually decreasing in value moving from EBITDA to EBIT, but it still remains positive observing the net profitability, with the exception of 2019, the year of the COVID pandemic. However, as shown in Fig. 4.17, the incidence of the spread is around 70% for all indexes used.

The countries with national WRA show a stable trend of EBIT, which is the margin coming from the regulated items (OPEX and CAPEX); while, moving to net profit, a high instability features both clusters: for "national" this could derive from the annual dynamic of the cost of debt, which cannot be immediately intercepted by the WRA with the rules designed to estimate allowed revenues, which cannot be immediately updated to the new marked conditions.

The integrated water sector has a higher net profitability rate (7.37) comparatively with only water utilities (4.61).

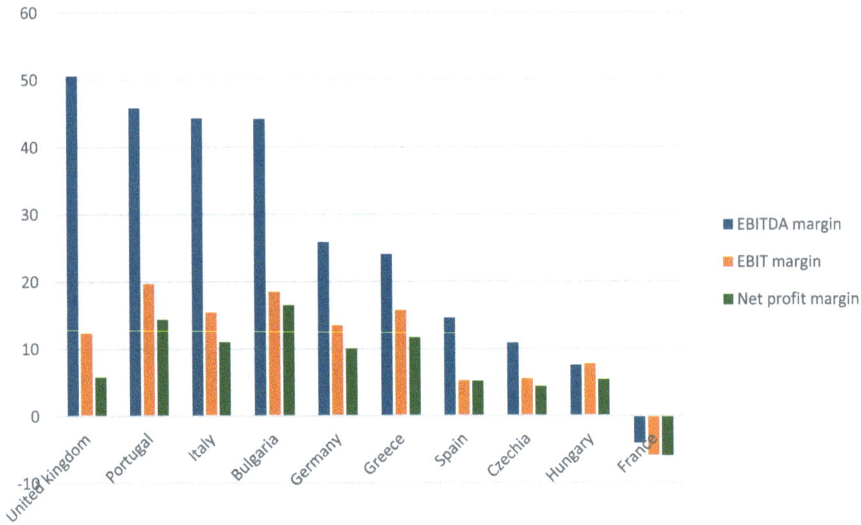

Fig. 4.14 Profitability indexes per country

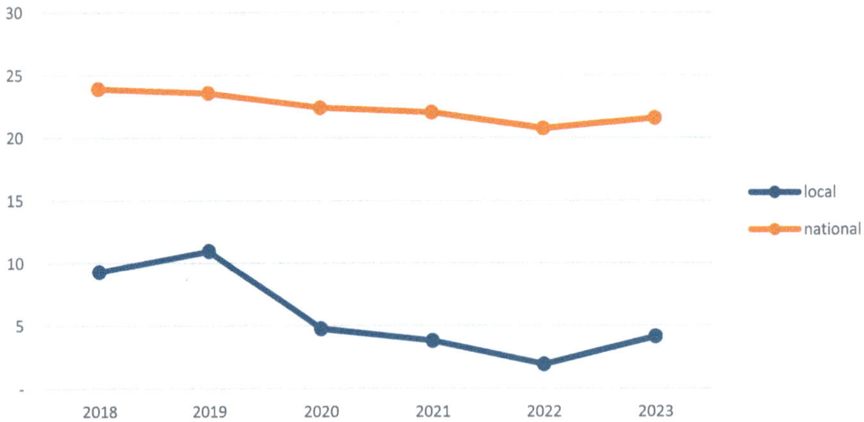

Fig. 4.15 EBIT margin trend during the observed period per regulatory model

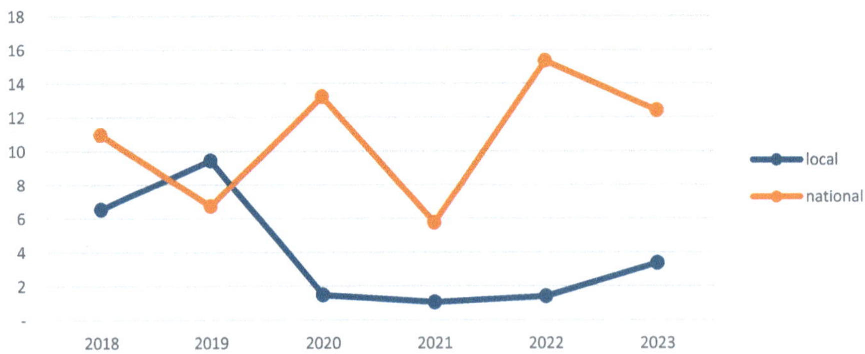

Fig. 4.16 Net profit margin trend during the observed period per regulatory model

Index	National	Local	Variance	Variance %
EBITDA margin	44.04	11.26	32.77	74%
EBIT margin	15.67	3.99	11.67	74%
Net profit margin	10.72	2.98	7.73	72%

Fig. 4.17 Spread of profitability per regulatory model

4.6 Operational Efficiency

The operational efficiency represents one of the main leading factor of profitability, linked to the production function and aiming to produce more with the same amount of resources, according to a output oriented perspective. The next figures in this section analyze the items of VoP, identified in: OPEX, Depreciation and Amortization (DA), and EBIT margin. All these items are related to the number of inhabitants served by each utility, considered as an output.

Figures 4.18 and 4.19 show the composition of VoP for "national" and "local" regulatory model: the amounts of OPEX per inhabitant are quite similar, and the EBIT margin is as great as the DA in both clusters, with the highest figures under a national authority due to the largest amount of investment realized. At a first glance the average values per cluster would suggest that national WRAs would be more oriented to incentivize the realization of investments, but few benefits deriving from regulation may be appreciated in terms of operational efficiency. This insight is confirmed observing the trend of OPEX per inhabitant of Fig. 4.20: the "national" cluster shows a smoother slope, avoiding the peak achieved in 2022 by the other cluster, but this gap is partially recovered in 2023, with an increase of cost per capita. If the presence of a WRA is not assuring higher firms' efficiency, it seems to grant predictability and cost stability.

However, a similar OPEXpc in both clusters cannot be easily justified. From one side, the high attention of WRA on outputs could at least partially explains a certain amount of the OPEX incurred by the regulated entities, for achieving the targets assigned (e.g. leakeages monitoring and reduction); but, from another side, the highest rate of asset renewal associated with this cluster could make viable the achievement of cost efficiencies.

A country perspective allows to appreciate the dynamic of VoP items: United Kingdom, Italy, Germany and Portugal record the highest spread between VoP and OPEX + DA, thus showing the highest operative gross profit per inhabitant or, in other words, the highest amount paid by each customer to cover the utility financial costs (Fig. 4.21).

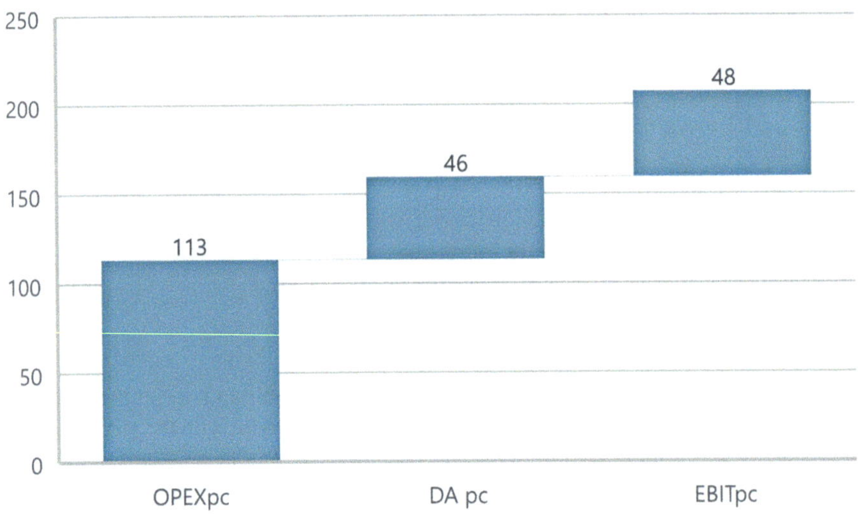

Fig. 4.18 The composition of VoP per inhabitant for the national regulatory cluster

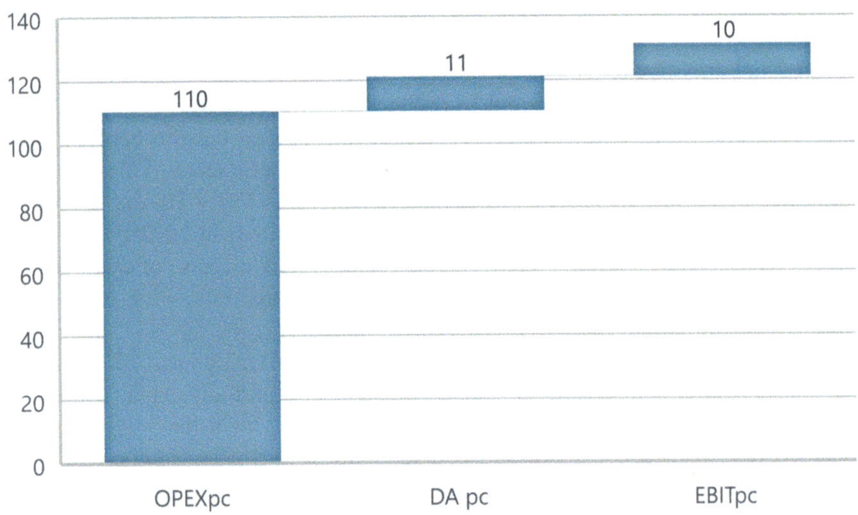

Fig. 4.19 The composition of VoP per inhabitant for the local regulatory cluster

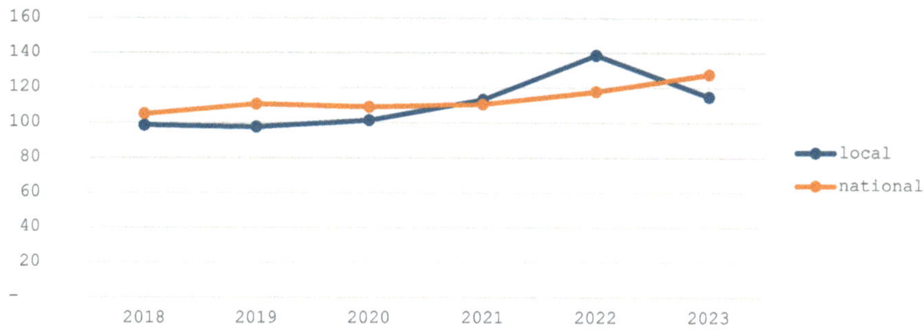

Fig. 4.20 OPEX per inhabitant trend during the observed period per regulatory model

Fig. 4.21 OPEX + DA per inhabitant vs. VoP per inhabitant per country

4.7 Turnover

The number of times the TA is renovated in a year through revenues (turnover) is an index which could provide some information on the investment strategy pursued by a company, since the higher the turnover the lower the amount of investments. This is the case for France and Hungary. However, while the latter achieve the highest rate of return of the panel thanks to a high turnover and a positive profitability (ROA = VoP/TA × EBIT/VoP), in the French utilities a negative profitability associated with a high turnover determines a set of negative rates.

In general, companies with a high turnover have a productive function mainly made of factors generating current expenditure (OPEX): this could be due to the license features, which, in some cases, does not include the asset renovation within the range of tasks of the utility (Affermage agreements); or alternatively the lack of investment can derive from a loose regulatory power exerted by public authorities.

Coming to the comparison between different regulatory models, the turnover is lower and more stable when a WRA has been set at national level, with an index ranged between 0.40 and 0.50 (Figs. 4.22 and 4.23).

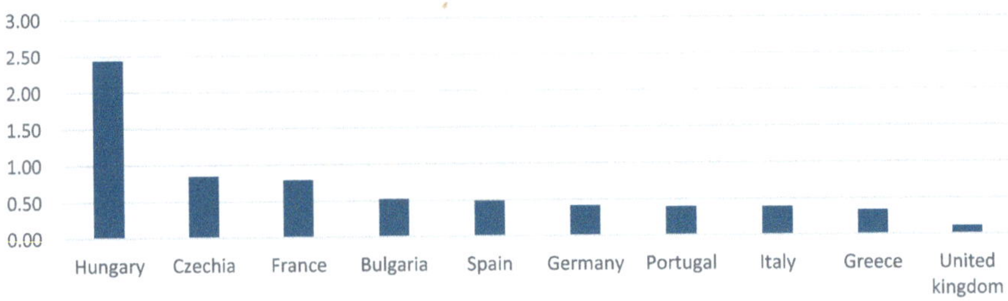

Fig. 4.22 Turnover index per country

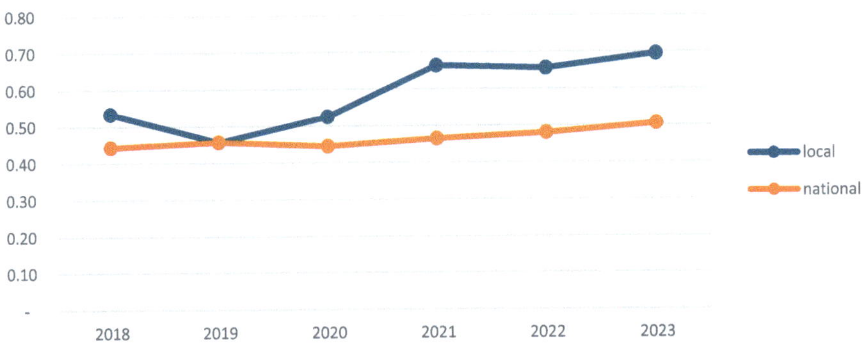

Fig. 4.23 Turnover trend during the observed period per regulatory model

Chapter 5
Discussion and Implications of the Study

The report analyses the performance of a panel of private owned utilities operating in Europe and serving a relevant quote of EU27 + England and Wales inhabitant (10%).

The study has multiple aims:

- providing a general overview of the performance of private owned utilities in the European water sector;
- identifying the attractiveness of each selected country, where private investors operate;
- analysing the effects of institutional country features, as the regulatory model;
- identifying the action exerted by some performance drivers.

Moving from the first goal, the water sector in Europe shows a positive performance, with an average ROE of 8.6% and a ROA of 4.16% for the panel, and a value of total asset of 68.8 bn. of euros, with an average of 485 € per inhabitant. A detailed analysis reveals significant variability, with varying trends across different indexes measured among countries. This explains the differing levels of attractiveness among countries, despite positive average data according to a global perspective.

Figure 5.1 provides several country rankings according to a selection of the main indexes adopted for the study: green, yellow and red color provide in a glance the attractiveness according to each single index; the color of the the first column of country names gives information about the overall attractiveness of the country.

Following an alphabetical order, **Bulgaria** offers a good return on investment, thanks to a high rate of profitability; the return on equity could be improved acting on the gearing, which is a lever not actually used, maybe for a low bankability also due to a low water tariff.

Czechia has a quite high tariff, especially if compared to other eastern countries; it could appear too high in comparison to the investment realized; the cost recovery measured through profitability is poor, as well as ROA, despite a significantly high

© The Author(s), under exclusive license to Springer Nature Switzerland AG 2026

A. Guerrini, *Private Investments in the European Water Sector*, SpringerBriefs in Water Science and Technology, https://doi.org/10.1007/978-3-032-24735-3_5

	Tariff	Investment	Gearing	ROE	ROA	Profitability	Turnover
Bulgaria	9	5	7	4	2	2	4
Czechia	3	6	8	8	7	9	2
France	6	8	9	10	10	10	3
Germany	2	3	4	5	4	5	6
Greece	10	9	6	7	6	3	9
Hungary	8	10	10	1	1	7	1
Italy	4	1	3	3	5	4	8
Portugal	5	4	2	2	3	1	7
Spain	7	7	5	6	8	8	5
United kingdom	1	2	1	9	9	6	10

Fig. 5.1 Country attractiveness for private investors

turnover. The country could have room for efficiency recovery, since it moves from the highest OPEXpc of the panel.

France shows some weaknesses within the panel. The level of full cost recovery is really poor and, despite a high turnover, this country presents negative rates of return, beside a wide variability with which is associated a high level of uncertainty. The level of investment per inhabitant is low.

Germany has a high level of investment financed with one of the highest tariff within the panel; profitability and rates of return are acceptable.

Greece has recorded a very poor level of investment, associated with the lowest water expenditure among the companies of the panel. Despite a good trend in profitability, a low turnover and gearing generate a poor return on equity.

The selected company operating in **Hungary** reaches the highest rates of return; this goal is achieved with an unusual turnover for the water sector, assured by a very low level of investment realized; also the tariff is low but it allows an overcompensation of costs. Despite the high ROE, the low room of investment realization impedes to this country to be the best where operate.

Italy and **Portugal** appear the best countries to invest in. **Italy** has the highest level of investment realization in the last year, despite a tariff growth quite contained; the return for shareholders is good (11.66%), also thanks to an effective usage of gearing; ROA and profitability are capped but are really stable, even during the Covid period, under the control of the national WRA. **Portugal** is partially acting on different factors, as Figure 4.21 shows: the relative lower level of investment per inhabitant seems to allow the achievement of higher returns (ROE at 22%) than Italy. The turnover is low in both countries, but it physiologically depends on the relevant amount of investment, comparatively to the rest of the panel.

Spain, as other selected countries, presents a quite relevant country risk for investors, with poor profitability and rates of return, also derived by a low tariff. The level of investment realized per inhabitant is poor.

The water sector in **United Kingdom** has a long history; today the country shows the highest tariff within the panel and a high level of investment; the high degree of indebtedness achieved in recent years by water utilities for boosting the ROE has generated a poor financial resilience, with relevant financial costs also due to do

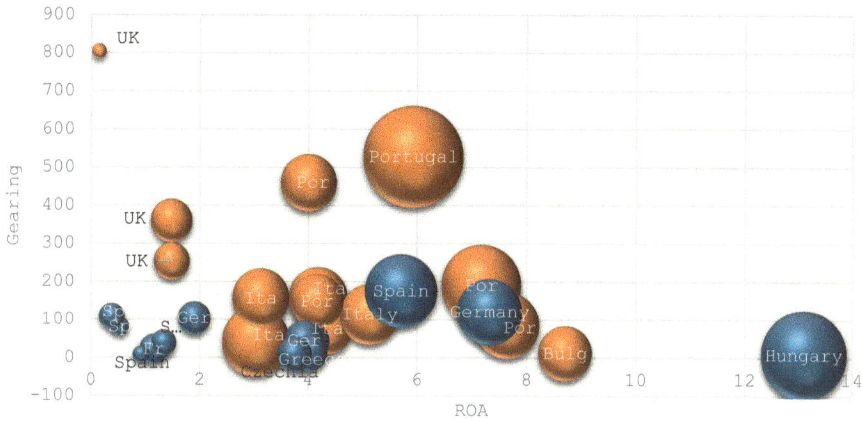

Fig. 5.2 The drivers of ROE

index linked bonds. Despite the full cost recovery is in place, measured by positive track record of profitability indexes, the more than proportional amount of investment realized in the last years of the period observed, determines low operative and net returns, meaning that this country is not over remunerating its water utilities.

The regulatory model seems to really affect the performance of private water utilities, as described in chapter three: the presence of a national WRA is related to highest rate of return, driven by profitability and gearing; then, the presence of an independent and specialized authority is a prerequisite for the investment realization: the cluster with a national WRA has a track record of investment per inhabitant 5 times the other companies.

The institution of clear, accepted and stable economic rules is a requisite for assuring companies' full cost recovery, granting financial resilience and bankability and, consequently, improving the condition for investments realization.

Considering Fig. 5.1, all countries with a WRA show a potential positive appeal for private investors, even if with different degrees. The only country with a different institutional setting to be included in the high ranked countries is Germany: yellow and red colored rows include only countries without a national regulator during the period monitored.[1]

Figure 5.2 is a chart with bubbles. The bubble's size measures the ROE, while x-axis and-y axis measures ROA and gearing, respectively. The orange bubbles represent companies of countries with national WRA, while the blue bubbles are the other companies.

Following the chart, ROE is driven by the "dash for debt" and by the return on investment: some countries (as Portugal) are acting on both; others are mainly focused alternatively on gearing (United Kingdom) or ROA (Hungary).[2] Considering

[1] Greece entrusted the independent energy authority only in 2023.

[2] Two French companies and a British company with negative ROE are not included in the chart.

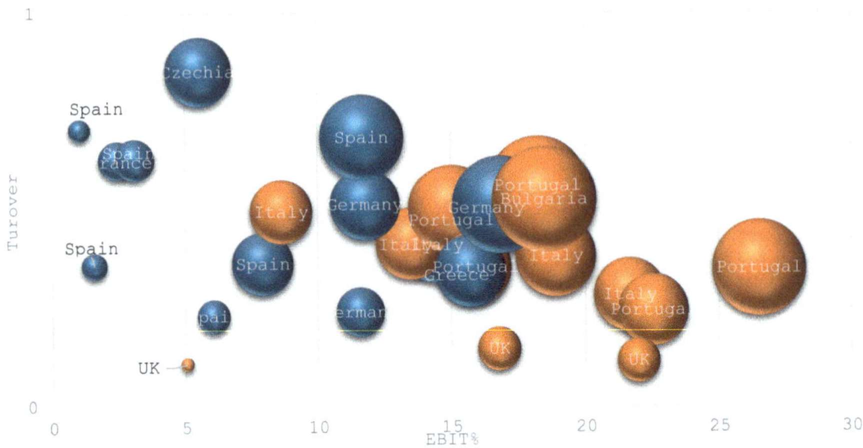

Fig. 5.3 The drivers of ROA

the regulatory framework, those companies operating under the control of a WRA show higher return for shareholders, with higher ROA (if the Hungarian outlier is removed), and gearing.

Moving to the analysis of the factors affecting ROA, Fig. 5.3 represents all the companies, with the exception of those with negative performance and the Hungarian outlier. The bubble's size measures the ROA, while the x-axis and y-axis measures the ratio EBIT to VoP and the Turnover, respectively. The figure shows quite clearly the different pattern of performance for the two cluster examined: the group of companies with WRA records high profitability and low turnover; while the contrary occurs for the countries without regulation or with locally based regulation.

As described in Chap. 3, grouping companies according to their scope of action, the integrated water service achieves the highest value for water expenditure, investment realization per inhabitant, profitability and gearing, while the pure water utility group gets the highest turnover and rates of return for shareholders.

Finally, in order to enquire any possible effects of firm's size on OPEX, Fig. 5.4 relates the cost per capita—without DA—with the scale of each utility, measured in terms of inhabitants. According to prior literature, the data show economies of scale within a certain threshold, which in the current study could be identified around 3 mn of inhabitants served (a large water utility); while for those very large companies some negative effects deriving from a too big scale of operation seem to rise.

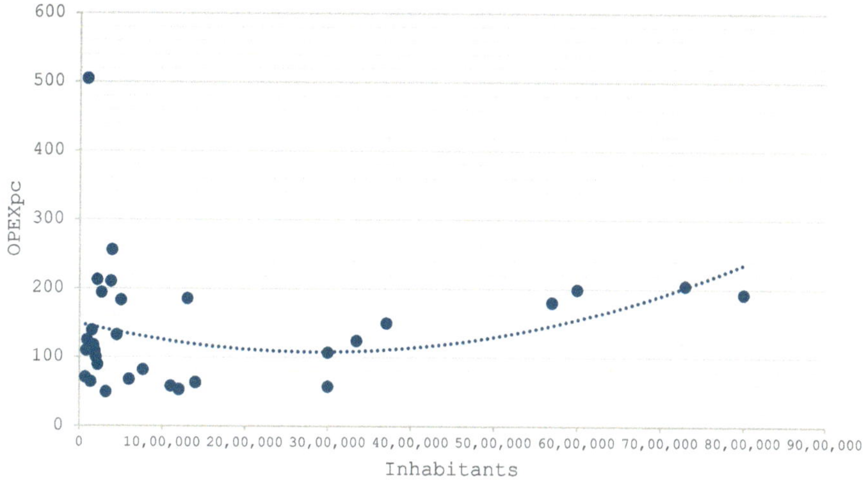

Fig. 5.4 The impact of utilities' scale on cost efficiency

Chapter 6
Final Remarks

The country profile is not neutral for private investors when they evaluate any opportunity of businesses in the water sector.

This report shows a different attractiveness of the industry among the selected European countries. The main institutional variable monitored is the presence of a WRA empowered for overseeing the sector and setting the framework of rules concerning cost allowance, quality and tariff structure.

The entrustment of regulatory power to a WRA assures high stability and predictability of revenues collected by water utilities, allowing the drafting of credible financial plan which have higher likelihood to be financed by third parties, as financial institutions. Beside the low variance of financial results, the presence of a WRA assures high margins for operators and, consequently, high rate of returns on investments.

According to the evidences of the current study, the European WRA seems to act following an objective function of investment maximization, in order to cover the large infrastructural needs of the sector and to overpass the poor quality of services delivered in certain areas of the countries. This goal is actually achieved through adequate profit margins, which assure the coverage of operating and financial costs and generate positive return on assets; then, the return for shareholders (ROE) is further improved with the increase of financial leverage by water companies, to earn from the differences among the notional financial costs and the actual cost of debt effectively paid.

This scheme has actually boosted investments in the last decade in many countries, generating some positive benefits also in terms of quality, as recorded by WRAs documents.[1]

[1]1 See for instance (ARERA 2025), Sintesi della Relazione Annuale, pp. 93–94, https://www.arera.it/chi-siamo/relazione-annuale/relazione-annuale-2025

© The Author(s), under exclusive license to Springer Nature Switzerland AG 2026

A. Guerrini, *Private Investments in the European Water Sector*, SpringerBriefs in Water Science and Technology, https://doi.org/10.1007/978-3-032-24735-3_6

Despite these positive results, the objective function mainly adopted by the European WRAs could generate some flaws which could impact on customers or, in the long term, on water utilities too.

A possible negative implication is the Averch-Johnson effect, associated to the rate of return regulatory models, which incentivize the investments realization "at any cost", limiting the policies for cost savings and the promotion of efficiency.

As a matter of fact, efficiency seems to be the missing goal which water stakeholders are expecting from the WRA activities. Despite this aspect would require further analysis, it is quite surprising that the cluster of countries with a WRA achieved a similar cost per inhabitant than those without the regulator.

Finally, the rules commonly used for WACC estimation risk to generate a negative loop, moving the water utilities toward a pattern of unsustainable debt growth.

As described in the Chap. 2, WRAs are implementing some adjustments to limit the described drawbacks, recurring to: Total Expenditure tariff models (TOTEX); incentive mechanisms for efficiency improvement, often based on benchmarking and cost frontiers; new procedures for gearing estimation.

This evolutionary pattern is creating the right conditions for continuing to paving the road for investments realization at efficient cost.

As last issues observed, scale and scope of action are variables positively acting on financial results, with some thresholds rising when firm's size is observed.

The current report may serve as an initial step for the development of a monitoring platform of the performance of private owned water utilities, which could be useful for:

- international institutions, to evaluate how water governance features are affecting the attractiveness of the sector;
- private investors, to identify those countries where it is more convenient to operate;
- water utilities, to compare their performance with an international panel of comparables.

The platform may incorporate tools for a comprehensive comparative analysis of concession frameworks and contract schemes across various countries to assess the risk of incompleteness, the generation of stranded costs, and challenges in initiating renegotiations. Then, a monitoring system for tenders could be activated in water utilities offices, in order to concretely evaluate any opportunities of entrance in a country.

Reference

ARERA (2025) Relazione Annuale. www.arera.it

Annexures

1.1 Annex I. General Data

See Tables A.1 and A.2

1.2 Annex II. Country Profile[1]

1.2.1 Bulgaria

	Bulgaria	data source
Area (sq km)	110,994	Wareg
Population (mn)	7.1	Wareg
Density (inhab/sq km)	64	Wareg
Per capita consumption (lit/inhab/day)	101	Eurostat
Operators	45	Wareg
Turnover (mn)	431.5	Wareg
Employees	16,000	Wareg
Fresh water per 1,000 inhabitants	15	Eurostat
Lenght of water network (1,000 km)	73.40	Eurostat
Sludge (1,000 ton)	44.00	Eurostat

[1] The country profile of the following countries have been taken from WAREG, Water Regulatory Governance across Europe: Bulgaria, Hungary, Italy, Portugal and Spain.

© The Editor(s) (if applicable) and The Author(s), under exclusive license to Springer Nature Switzerland AG 2026

A. Guerrini, *Private Investments in the European Water Sector*, SpringerBriefs in Water Science and Technology, https://doi.org/10.1007/978-3-032-24735-3

Table A.1 Data of 2023 for the Water collection. Treatment and supply sector (Eurostat)

Country	Number of enterprises	Persons employed	Persons employed per enterprise	Turnover mn euros
European Union—27 countries (from 2020)	15.200	330.945	22	60.000.00
Belgium	49	7.603	155	3.275.71
Bulgaria	64	15.663	245	494.64
Czechia	851	18.285	21	2.603.99
Denmark	1.704	2.341	1	715.45
Germany	1.704	37.337	22	16.949.29
Estonia	70	1.199	17	159.79
Ireland	214	1.958	9	:
Greece	588	7.010	12	521.94
Spain	3.162	47.422	15	9.422.24
France	1.397	21.728	16	4.306.66
Croatia	111	7.008	63	294.31
Italy	747	36.460	49	10.555.87
Cyprus	6	414	69	148.18
Latvia	64	1.573	25	114.00
Lithuania	57	4.399	77	269.03
Luxembourg	6	189	32	89.97
Hungary	164	15.763	96	610.44
Malta	35	1.149	33	:
Netherlands	38	5.976	157	1.898.84
Austria	1.055	1.812	2	597.25
Poland	785	27.761	35	1.792.37
Portugal	171	14.972	88	1.905.50
Romania	367	34.899	95	1.409.98
Slovenia	82	4.820	59	652.92
Slovakia	70	7.998	114	547.47
Finland	1.329	2.379	2	1.199.15
Sweden	310	2.827	9	885.48
Iceland	11	7	1	0.52
Norway	633	278	0	56.46
Switzerland	110	1.812	16	748.12
Bosnia and Herzegovina	142	7.551	53	168.45
Montenegro	:	:	:	:
North Macedonia	71	5.474	77	86.87
Albania	:	:	:	:
Serbia	167	17.654	106	455.60

Table A.2 Data of 2023 for the Water collection. Treatment and supply sector (Eurostat)

Country	Number of enterprises	Persons employed	Persons employed per enterprise	Turnover mn euros
European Union—27 countries (from 2020)	12.685	154.000	12	:
Belgium	229	2.879	13	1.238.67
Bulgaria	55	382	7	17.28
Czechia	186	1.400	8	153.04
Denmark	300	1.808	6	1.599.76
Germany	1.924	46.687	24	11.046.31
Estonia	68	160	2	11.84
Ireland	217	1.486	7	2.455.34
Greece	463	1.333	3	92.58
Spain	599	6.807	11	828.76
France	1.425	10.521	7	1.893.90
Croatia	78	379	5	71.09
Italy	1.379	11.637	8	1.927.34
Cyprus	64	500	8	119.13
Latvia	51	1.314	26	:
Lithuania	49	591	12	26.70
Luxembourg	9	:	:	:
Hungary	511	5.753	11	315.51
Malta	6	9	2	:
Netherlands	129	2.342	18	415.09
Austria	1.002	3.907	4	1.142.49
Poland	2.440	42.492	17	3.076.91
Portugal	86	2.586	30	457.02
Romania	318	1.841	6	97.79
Slovenia	44	578	13	72.07
Slovakia	450	825	2	54.95
Finland	338	1.601	5	417.93
Sweden	265	4.159	16	913.99
Iceland	4	18	5	2.58
Norway	186	1.306	7	291.41
Switzerland	194	3.857	20	1.003.99
Bosnia and Herzegovina	10	148	15	3.25
Montenegro	:	:	:	:
North Macedonia	8	252	32	5.63
Albania	:	:	:	:
Serbia	:	:	:	:

1.2.1.0.1 The Regulator

The Bulgarian National Energy and Water Regulatory Commission (EWRC) was established in 1999 with competencies over the water sector since 2006. The EWRC is a multi-sector regulator, which is organised as a National WRA.

The EWRC is formed by a Commission that includes Chairman, Lawyer, Economist, 4 Energy experts, and 2 Water and sanitation experts. EWRC employs 166 people of those 26 people work for the Water and Sanitation Directorate. As the Regulatory Authority for water, EWRC main responsibilities are:

- to regulate the quality of water and sewerage services through Key Performance Indicators;
- to approve 5-years business plans and tariff of water and sanitation services;
- to review customer complaints;
- to approve the common terms and conditions of contracts for the provision of WSS services to consumers; and
- to control and impose sanctions.

1.2.1.0.2 Legal Framework

The main legislation on the water sector of Bulgaria is the following one:

- The Water Act regulates the integrated water resource management. Including the requirements for ownership, management, operation, and planning of public water and sanitation assets and the duties of the Water Associations (for the regional operators) and Local Municipalities (for the small municipal operators);
- The Act on Regulation of Water and Sanitation Services provides norms for the national regulation of tariffs and quality of water and sanitation services, provided by the Energy and Water Regulatory Commission (EWRC);
- The Energy Act provides norms for the organisation of work of EWRC; and
- Multiple by-laws concerning the water and sanitation sector are available.

1.2.1.0.3 Institutional Framework

The Ministry of Regional Development and Public Works is the sector ministry responsible for coordinating the water and sanitation sector management at the national level. It develops and implements the State water policy and acts as a principal owner of the regional companies owned by the state. The Ministry of Environment and Waters is responsible for environmental protection and water resource management. The Ministry of Health is responsible for monitoring drinking water quality. Water Associations act as public owners of the water and sanitation infrastructure when ownership of assets in the designated service area is split between the state and the municipalities or between several municipalities and local municipalities (when the ownership of the assets in the designated service area belongs only to the Municipality). Water Associations should plan their future development through regional master plans and investment programs, select operators. and conclude delegation contracts for asset operation and service provision.

1.2.1.1 Republic of Czech

	Republic of Czech	data source
Area (sq km)	78,871	wikipedia
Population (mn)	10.8	wikipedia
Density (inhab/sq km)	137	wikipedia
Per capita consumption (lit/inhab/day)	89	Eurostat
Operators	1,037	Eurostat
Turnover (mn)	2,757	Eurostat
Employees	19,685	Eurostat
Fresh water per 1,000 inhabitants	1.4	Eurostat
Lenght of water network (1,000 km)	78.80	Eurostat
Sludge (1,000 ton)	235.00	Eurostat

1.2.1.1.1 Institutional Framework[2]

The supervision of the water management in the Czech Republic is surprisingly fragmented. It is mainly carried out by the Ministry of Finance and the Ministry of Agriculture, but the Ministry of the Environment, the Ministry of Health, the Ministry of Transport and the Ministry of Defense also have partial competences. Every year, the Ministry of Finance oversees the creation of the prices of drinking and waste water. The Ministry of Agriculture ensures the regulation of water companies, supervises compliance with the binding rules defined in agreement with the Ministry of Finance, collects data on the costs of water and wastewater companies and carries out inspections of WWS. It ensures that customers know who to contact with complaints about water companies when a support is needed. The Department of Water Protection at the Ministry of the Environment covers the protection of the quantity and quality of surface and underground water, protection against floods, planning in the field of water at the national and international level, international cooperation in the field of water protection, economic, financial and administrative instruments for water protection, the creation of legislation and standards in the field of water protection.

The regulation of water management in the Czech Republic is extremely fragmented compared to other countries, as it is ensured by six ministries. In the Czech Republic, there is no independent professional regulatory body for water management, and the regulation of network sectors has long been perceived in the country as the regulation of electricity, gas and heat.

---

[2]Zimkovà et al., 2023. Efficiency Evaluation of Water Sector in the Czech Republic: Two-Stage Network Dea, *Statistika, 103(4): 462–475.*

1.2.1.2 France

	France	data source
Area (sq km)	551,695	wikipedia
Population (mn)	64.9	wikipedia
Density (inhab/sq km)	118	wikipedia
Per capita consumption (lit/inhab/day)	165	Eurostat
Operators	2,822	Eurostat
Turnover (mn)	6,201	Eurostat
Employees	32,249	Eurostat
Fresh water per 1,000 inhabitants	3	Eurostat
Lenght of water network (1,000 km)	906.00	Eurostat
Sludge (1,000 ton)	1,174.00	Eurostat

1.2.1.2.1 Institutional Framework[3]

The central government is responsible for:

- the development and implementation of water legislation;
- developing national water policy and addressing common issues across river basins;
- coordination of cooperation on water management with neighbouring countries and at international level.

Regional councils are represented on the river basin committees (*comités de bassin*), which set river basin objectives, adopt the RBMPs and FRMPs and provide coordination among elected bodies and stakeholders. The prefect coordinators of the river basins give final approval to the RBMPs adopted by the river basin committees. They coordinate water management in their river basins, including enforcement of water regulations. The Regional directorates for the environment, planning and housing (DREAL) support the authorities on water management, including monitoring of water bodies. The water agencies provide technical and scientific support at river-basin level, including for the preparation of RBMPs. Moreover, they provide financial resources for implementing the Programmes of Measures.

Municipalities are responsible for:

- local implementation of water policies and plans, including local water management projects.
- managing drinking water supply, municipal sewerage and waste water treatment (either directly via municipally owned companies or delegated to private operators).
- local water protection and flood prevention measures.

Municipalities can joint together in inter-municipal unions (*Syndicat intercommunaux des eaux*) to manage common water services or undertake other water management activities. Municipalities and inter-municipal unions are represented on the

[3] https://portal.cor.europa.eu/divisionpowers/Pages/France-Water-Management.aspx

river basin committees. Municipalities participate in Local Water Commissions (*Commission locale de l'eau*) together with water users and stakeholders. These Commissions develop local water plans. (*Schéma d'Aménagement et de Gestion des Eaux*). Municipalities and inter-municipality groups participate in river contracts that implement measures for the RBMPs, FRMPs and lower-level plans.

1.2.1.3 Germany

	Germany	data source
Area (sq km)	357,592	wikipedia
Population (mn)	83.28	wikipedia
Density (inhab/sq km)	233	wikipedia
Per capita consumption (lit/inhab/day)	126	Eurostat
Operators	3,628	Eurostat
Turnover (mn)	27.996	Eurostat
Employees	84.024	Eurostat
Fresh water per 1.000 inhabitants	2	Eurostat
Lenght of water network (1.000 km)	544.00	Eurostat
Sludge (1.000 ton)	1,750.00	Eurostat

1.2.1.3.1 Institutional framework[4]

Central Government:

- formulates recommendations in the form of guidelines for the implementation of the WFD and the FD by the Länder (through a Working Group);
- acts as a fora for Länder on transboundary issues;
- updates and optimizes the scientific bases and measures for safe drinking water;
- for some river basins, a common inter-land RBMP (B-level plan) is prepared through collaboration of more than one Länder. Implementation remains with the competent Länder.

Regional Goverment;

- coordination and strategic direction of water management at lower levels;
- preparing the RBMP and the FRMP;
- licencing polluting industry installations;
- public participation;
- reporting and measurement.

Intermediate level:

- water resource planning;
- household waste collection and disposal;
- permitting;
- monitoring of rivers.

[4] https://portal.cor.europa.eu/divisionpowers/Pages/Germany-Water-Management.aspx

Local Authorities:

- procedures under water management law;
- monitoring of (smaller) water bodies and discharges;
- providing expert advice;
- maintenance and management of sewerage and waste water treatment;
- drinking water supply.

1.2.1.4 Greece

	Greece	data source
Area (sq km)	131,957	Wareg
Population (mn)	10.77	Wareg
Density (inhab/sq km)	82	Wareg
Per capita consumption (lit/inhab/day)	139	Eurostat
Operators	1,051	Eurostat
Turnover (mn)	615	Eurostat
Employees	8,343	Eurostat
Fresh water per 1,000 inhabitants	7	Eurostat
Lenght of water network (1,000 km)	76.50	Eurostat
Sludge (1,000 ton)	99	Eurostat

1.2.1.4.1 Institutional Framework[5]

Central Goverment:

The Special Secretariat for Water (SSW) under the Ministry of Environment and Energy are responsible for:

- coordinating water management issues;
- implementing the WFD;
- monitoring the quality and quantity of water;
- overseeing and regulating waste water and reuse and flood management;
- engaging the public;
- approving all regional RBMPs and FRMPs.

The National Council for Water (NCW) is responsible for:

- developing the national strategy on the management and protection of Greek waters;
- approving the national RBMP and FRMP prepared by the SSW.

Regional Government.

Regional authorities are responsible for:

- licensing discharges of industrial wastewater and municipal wastewater from treatment plants.

[5] https://www.raaey.gr/energeia/en/about/

The Regional Water Departments are responsible for:

- overseeing or preparing the preparation the RBMPs and the FRMPs in their Region. The Regional Water Departments can transfer that competence to the Special Secretariat for Water, which was the case in the previous cycle of implementation for all but two Regions;
- engaging the public in the preparation of the RBMP and FRMP.

Local Government.

The Municipalities are responsible for:

- participating in public consultations for the preparation of the RBMPs and FRMPs;
- protecting and managing water resources from extensive fisheries and pollution;
- constructing, maintaining and managing local water supply, irrigation, and sewage systems.

With Law 5037/2023 the former Authority for Energy was renamed as Regulatory Authority for Energy, Waste and Water (RAAEY) and its scope is expanded with responsibilities over Water services and Urban Waste management. Specifically, the Authority's responsibilities are defined in the amended article 4 of Law 4001/2011: "The control, regulation and supervision of the energy market , well as the control, regulation and supervision of subjects related to the provision of water services and urban waste management, without prejudice to the competences of the Minister of Environment and Energy, are exercised by the Regulatory Authority of Energy. Waste and Water. which is the national regulatory authority in matters of electricity and natural gas. water services and urban waste management, within the meaning of Directives (EU) 2019/944 and 2009/73/EC and Laws 3199 /2003 (A' 280) and 4819/2021 (A' 129)." It is noted that the operation of the Waste and Water Departments begins with the appointment of its Members and the new responsibilities are assumed after the approval of the new Organization and the Regulation of administrative operation and management of the Authority, as well as with the completion of the necessary hiring procedures, which are determined by a decision of the Minister of Environment and Energy, published in the Government Gazette, following the opinion of the Authority's Board. In accordance with the provisions of Law 4001/2011, RAAEY is legally independent. While possessing administrative and financial independence. and is subject only to parliamentary and judicial control (Article 5 and Article 6 par. 1).

1.2.1.5 Hungary

	Hungary	data source
Area (sq km)	93,023	Wareg
Population (mn)	9.77	Wareg
Density (inhab/sq km)	105	Wareg
Per capita consumption (lit/inhab/day)	101	Wareg
Operators	40	Wareg
Turnover (mn)	740	Wareg
Employees	18,881	Wareg
Fresh water per 1,000 inhabitants	n.a.	Eurostat
Lenght of water network (1,000 km)	68.40	Eurostat
Sludge (1,000 ton)	226	Eurostat

1.2.1.5.1 The Regulator

The Hungarian Energy and Public Utility Regulatory Authority (HEA) is a multi-sector independent WRA established in 2013 that employs 330 people (60 in the water area) with an annual budget of 31 mn €.

As WRA, HEA main responsibilities are:

- Price supervision. The Authority is entitled to supervise, control and regulate the prices of water utility service;
- Submitting a proposal on prices (tariff setting). The Authority is entitled to submit a proposal on the utility tariffs each year to the minister responsible for the water utility supply;
- Licensing powers. Supplying drinking water and managing wastewater can only operate if thery are in possession of the license granted by the Authority. The Authority also has the right of granting the application of prices differing from the utility tariff;
- Approving Authority. It is also the right of the Authority to approve the "rolling development plans" (15 years development plans), which consist of development, replacement and investment design plans.

1.2.1.5.2 Legal Framework

The main legislation on the water sector of Hungary is the following one:

- The Act CCIX of 2011 on Water Utility Supply;
- The Government Decree 58/2013. (II. 27.) on the implementation of the Act CCIX of 2011 on Water Utility Supply (58/2013. (II. 27.); and
- Act XXII of 2013 on the Hungarian Energy and Public Utility Regulatory Authority.

1.2.1.5.3 Institutional Framework

The water sector in Hungary is managed at the National and Regional levels.

The Hungarian Energy and Public Utility Regulatory Authority (HEA) is the WRA responsible for the economic regulation. Licensing and monitoring of sector players.

The General Directorate of Water Management and its regional directorates are responsible for the management and protection of water resources and the governance of administrative water authorities. Altogether water resources management. These entities are part of the Ministry of the Interior.

The National Directorate General for Disaster Management and its 12 regional directorates are responsible for water protection. Licensing and registration of water facilities. These entities are part of the Ministry of the Interior.

The National Public Health and Medical Officer Service is responsible for drinking water and bathing water quality regarding public health. The Service is under the supervision of the Minister of Human Resources.

Municipalities are responsible for ensuring local water utility services to service users in their territory. Most municipalities are also the owners of the water utility infrastructure. While the five state-owned regional water utilities own a part of the country assets.

1.2.1.6 Italy

	Italy	data source
Area (sq km)	301,338	Wareg
Population (mn)	60.59	Wareg
Density (inhab/sq km)	201	Wareg
Per capita consumption (lit/inhab/day)	204	Wareg
Operators	2,126	Eurostat
Turnover (mn)	12,483	Eurostat
Employees	48,097	Eurostat
Fresh water per 1,000 inhabitants	n.a.	Eurostat
Lenght of water network (1,000 km)	474	Eurostat
Sludge (1,000 ton)	n.a.	Eurostat

1.2.1.6.1 The Regulator

The Italian regulatory Authority for Energy. Networks and Environment (ARERA) is the independent multi-sector regulator for the whole country. It employs approximately 250 persons. Its revenues come only from regulated entities. in 2020 they amounted to 68.95 mn €. Its powers to regulate the integrated water services were provided by law in 2011.

As the Regulatory Authority for water, ARERA main responsibilities are:

- setting tariffs for protected markets and the use of monopoly infrastructures (define cost components. Set/update tariff methodology for tariff determination. Approve tariffs);
- setting compulsory standards for technical and commercial quality of service, through incentives, penalties, refunds;
- setting rules for administrative and accounting separation of vertically integrated companies;
- setting minimum conditions to be included in investment programmes and economic-financial plans provided by Local Authorities. and monitor their compliance;
- collecting economic-financial data from operators, ensure compliance of Regulation;
- consumer protection, infringement procedures, dispute settlement; and
- providing opinions to Government and Parliament.

1.2.1.6.2 Legal Framework

The main legislation on the water sector of Italy is the following one:

- Law decree 133/2014. The general rules for the organisation of integrated water services and the governance of the sector;
- Resolution 639/2023/R/idr. The Water Tariff Method;
- Resolution 656/2015/R/idr. The standard agreement for the Regulation of relationships between entrusting entities and providers of the water services;
- Resolution 917/2017/R/idr. Regulation of technical quality; and
- Resolution 186/2020/R/idr. Contractual quality of water services.

1.2.1.6.3 Institutional Framework

The water sector in Italy is managed at the national, regional and sub-regional levels. The Ministry of Environment defines general principles of water uses and services at the national level. Provides specific rules, mechanisms and procedures concerning the integrated water sector and ensures its compliance. Regions set the geographical bounds of the area into which integrated water service has to be entrusted, called "ATO" or "Ambito Territoriale Ottimale". In addition, they identify the Local Authority – EGA, in charge of organising the integrated water service at the local level, planning investments and entrusting the management of service to the operator.

1.2.1.7 Portugal

	Portugal	data source
Area (sq km)	92391	Wareg
Population (mn)	10.31	Wareg
Density (inhab/sq km)	112	Wareg
Per capita consumption (lit/inhab/day)	189	Wareg
Operators	535	Wareg
Turnover (mn)	2.035	Wareg
Employees	31.414	Wareg
Fresh water per 1.000 inhabitants	7	Eurostat
Lenght of water network (1.000 km)	112.40	Eurostat
Sludge (1.000 ton)	119	Eurostat

1.2.1.7.1 The Regulator

The Water and Waste Services Regulation Authority (ERSAR) is the WRA that executes essential regulatory functions on water supply and sanitation of urban wastewater in mainland Portugal. It was established in 2009 and, since 2014, acts independently from the Government. It employs approximately a hundred of people with an annual Budget of 10.3 mn. €.

As a WRA, ERSAR main responsibilities are:

- to ensure the quality of the services rendered by drinking water supply systems, urban wastewater and municipal waste systems;
- to supervise the creation. Execution. management and operation of those systems;
- to protect consumer rights;
- to safeguard the sustainability and economic viability of the municipal and regional water and waste utilities;
- to monitor and control the drinking water quality for all the operators in mainland Portugal;
- to coordinate and supervise the application of the Drinking Water Directive (Council Directive 98/83/EC);
- to coordinate the task to transpose the new Drinking Water Directive (Directive EU 2020/2184) to national legislation;
- to issue opinions about tariffs for the Ministry of Environment and Climatic Action (state-owned systems) and the Municipalities (municipal-owned systems);
- to issue recommendations;
- to approve regulations with external effectiveness;
- to issue sanctions; and
- to collaborate with the Government in legal developments and strategy definition for the water sector.

1.2.1.7.2 Legal Framework

The main legislation on the water sector of Portugal is the following one:

- Law 23/96. amended by law 51/2019, Law 10/2013, Law 44/2011, Law 6/2011, Law 24/2008 and Law 12/2008 (essential public services);
- Decree-Law 306/2007 amended by Decree-Law 152/2017 (water quality for human consumption);
- Decree-Law 194/2009 amended by Decree-Law 92/2010 and Decree-Law 12/2014 (municipal services for the public water supply. Wastewater sanitation and urban waste management); Decree-Law 92/2013. amended by Decree-Law 72/2016 and Decree-Law 16/2021 (multi-municipal services for the public water supply. Wastewater sanitation and urban waste management);
- Law 67/2013 amended by law 12/2017. law 71/2018 and law 75-B/2020 (Framework Law for regulatory bodies);
- Decree-Law 10/2014 (Statutes of the Water and Waste Services Regulation Authority);
- Decree-Law 96/2014 (Granting the management of multi-municipal systems for the treatment and selective collection of municipal waste to exclusively or primarily private capital entities); and
- Decree-Law 102-D/2020, 10 December (new general waste management framework and a new landfill of waste framework).

1.2.1.7.3 Institutional Framework

The Parliament of the Portuguese Republic approves laws that proceed for President of the Republic promulgation.

The State Government approves decree-laws. Resolutions. ordinances and deliberations.

The Ministry of Environment and Climatic Action approves tariffs and new investments not foreseen in the concession contract for state-owned systems; approves the sector's strategic plans.

Local Governments (Municipalities) approve tariffs and investments in municipal-owned systems (according to the autonomy principle of the local Government assigned by the Constitution).

1.2.1.8 Spain

	Spain	data source
Area (sq km)	505,944	Wareg
Population (mn)	47.45	Wareg
Density (inhab/sq km)	94	Wareg
Per capita consumption (lit/inhab/day)	128	Wareg
Operators	3,761	Eurostat
Turnover (mn)	10,251	Eurostat
Employees	54,229	Eurostat
Fresh water per 1,000 inhabitants	2	Eurostat
Lenght of water network (1,000 km)	248.20	Eurostat
Sludge (1,000 ton)	1,210	Eurostat

1.2.1.8.1 Legal Framework

The main legislation on the water sector of Spain is the following one:

- The Water Law (RDL 1/2001);
- The Local Regime Basis Law (Ley 7/1985);
- The Local Tax Law (RDL 2/2004);
- The Law on streamlining and sustainability of local administration (Ley 27/2013);
- Regional laws on wastewater treatment;
- Regional water acts;
- Local (Municipality) water supply service regulations; and
- Local or regional urban water services prices ordinances.

1.2.1.8.2 Institutional Framework

In Spain, Water services are a Municipal responsibility. Therefore, there is no Independent Regulatory Agency/Body/Organism at the National level. Although the water sector is highly regulated, it presents extraordinary fragmentation of competencies.

At the National level, the Ministry for the Ecological Transition and the Demographic Challenge (MITERD) has competencies in:

- hydrological planning;
- public water domain control; and
- granting concessions on water use.

1.2.1.9 England and Wales

	England and Wales	data source
Area (sq km)	151,058	Wikipedia
Population (mn)	60.24	Wikipedia
Density (inhab/sq km)	399	Wikipedia
Per capita consumption (lit/inhab/day)	142	Eurostat
Operators	27	Ofwat
Turnover (mn)	14,680	Eurostat
Employees	n.a.	
Fresh water per 1,000 inhabitants	n.a.	Eurostat
Lenght of water network (1,000 km)	416.90	Eurostat
Sludge (1,000 ton)	1,137	Eurostat

1.2.1.9.1 The Regulator[6]

The Water Services Regulation Authority, or Ofwat, is a non-ministerial government department and body responsible for the economic regulation of the privatised water and sewerage industry in England and Wales. Ofwat's main statutory duties

[6] https://en.wikipedia.org/wiki/Ofwat

include protecting the interests of consumers, securing the long-term resilience of water supply and wastewater systems, and ensuring that companies carry out their functions and are able to finance them.

Ofwat primarily sets limits on the prices charged for water and sewerage services, taking into account proposed capital investment schemes (such as building new wastewater treatment works) and expected operational efficiency gains. The most recent review was carried out in *2024*; reviews are carried out every five years.

The Water Act 2014 extended retail competition to all non-household customers of English water companies from April 2017 and provided for possible future competition in wholesale markets. Ofwat's role includes regulating such water and wastewater markets and promoting effective competition wherever appropriate.

1.2.1.9.2 Legal Framework[7]

The Water Act 1989 provided for the privatisation of the former water authorities. Water related legislation (including the Water Act 1989) was subsequently consolidated into new Acts of Parliament. These included:

- The Water Industry Act 1991, which sets out the main powers and duties of the water and sewerage companies, thus replacing those set out in the Water Act 1989, and defined the powers of the Director General of Water Services (now the Water Services Regulation Authority (Ofwat)).
- The Water Resources Act 1991, which set out the functions of the National Rivers Authority (now the Environment Agency) and introduced water quality classifications and objectives for the first time.

Subsequent Acts have modified the framework. These include the following:

- The Competition and Service (Utilities) Act 1992 increased Ofwat's powers to determine disputes and increased the limited opportunities for competition in the industry.
- The Environment Act 1995 led to restructuring of environmental regulation and placed a duty on the companies to promote the efficient use of water by customers. It created a new body, the Environment Agency, which took over the functions of the National Rivers Authority, Her Majesty's Inspectorate of Pollution, the waste regulation functions of local authorities and certain elements of the Department of the Environment. Natural Resources Wales now exercises the functions of the Environment Agency in Wales.
- The Competition Act 1998 prohibits any agreements between businesses that prevent, restrict or distort competition. It also prohibits any abuse of a dominant market position. Ofwat shares enforcement powers in relation to the water and sewerage sectors with the Competition and Markets Authority (CMA).

[7] https://www.ofwat.gov.uk/regulated-companies/ofwat-industry-overview/legislation/

- The Water Industry Act 1999 made several important amendments to the Water Industry Act 1991. These included:

  removing a company's right to disconnect domestic customers for non-payment of bills,
  limiting the circumstances in which companies can starting charging domestic customers on a metered basis,
  securing that companies could continue to charge customers on the basis of rateable value.

- The Water Act 2003 amended the framework for abstraction licensing, made changes to the corporate structure of economic regulation, and extended the scope for competition in the industry to large users.
- The Enterprise Act 2002 amended the Water Industry Act 1991 to update the regime for the compulsory reference of certain mergers between water companies to the Competition Commission (now the CMA).
- The Flood and Water Management Act 2010 encouraged the use of sustainable urban drainage systems (SUDs), amended the Water Industry Act to modernise the list of activities that can be restricted by water companies in a drought, and made it easier for water companies to offer lower tariffs to certain groups.
- The Water Act 2014 enables greater competition for non-household customers (expected to be limited to customers of English water companies) and gives Ofwat new powers to make rules about charges and charges schemes, as well as making provisions for flood insurance and drainage boards.

1.2.1.9.3 Institutional Framework[8]

Department for Environment, Foods and Rural Affairs:
Sets the overall water and sewerage policy framework in England. This includes:

- standard setting;
- drafting of legislation;
- creating special permits (e.g. drought orders).*Welsh Government:*

Sets the overall water and sewerage policy framework in Wales. This includes:

- standard setting;
- drafting of legislation;
- creating special permits (e.g. drought orders);

The Welsh Government and Defra work closely together on water issues.
Ofwat:
It is the WRA and act to:

[8] https://www.ofwat.gov.uk/regulated-companies/ofwat-industry-overview/

- protect the interests of consumers, wherever appropriate by promoting competition;
- make sure that the water companies properly carry out their functions;
- ensure that the water companies can finance their functions;

It also (among other things):

- promote economy and efficiency;
- contribute to the achievement of sustainable development;*Natural Resources Wales:*

The environmental regulator of the water and sewerage sector in Wales. Their purpose is to ensure that the natural resources of Wales are sustainably maintained, enhanced and used, now and in the future.*Environment Agency:*

The environmental regulator of the water and sewerage sector in England. They are the principal adviser to the government on the environment, and the leading public body protecting and improving the environment of England. They work in partnership with a range of other organisations to

- reduce flood risk
- promote sustainable development
- secure environmental and social benefits.*Drinking Water Inspectorate (DWI):*

The drinking water quality regulator. They check that the water companies in England and Wales supply water that is safe to drink and meets the standards set in the Water Quality Regulations. They do this by:

- checking the tests that water companies carry out on drinking water;
- inspect individual companies.

Consumer Council for Water (CCW):

They represent consumers within the water and sewerage sectors. They also investigate consumer complaints that have not been satisfactorily resolved by the water companies.

Natural England:

The government's advisor on the natural environment. They provide practical advice, grounded in science, on how best to safeguard England's natural wealth for the benefit of everyone. Their purpose is to protect and improve England's natural environment and encourage people to enjoy and get involved in their surroundings.

1.3 Annex III. The Materials Collected

See Tables C.1 and C.2.

Table C.1 The panel selected

Company	Country	Regulatory model	Main shareholder	Segment	Inhabitants
SOFIYSKA VODA	Bulgaria	national	Veolia	WWW	1,400,000
BRNENSKE VODARNY A KANALIZACE. A.S.	Czechia	local	Suez	WWW	380,000
SOCIETE DES EAUX DE L'OUEST PARISIEN	France	local	Suez	W	451,758
SOCIETE DES EAUX DE L AGGLOMERATION DU BASSIN DE BRIVE	France	local	Suez	WWW	81,386
VALYO	France	local	Veolia	W	92,667
SOCIETE DES EAUX DU PAYS DE MONTBELIARD	France	local	Veolia	WWW	137,055
SOCIETE MAHORAISE DES EAUX	France	local	Vinci	W	186,729
AVACON WASSER GMBH	Germany	local	E.ON	WWW	600,000
WASSERVERSORGUNGS-UND ABWASSERENTSORGUNGSGESELLSCHAFT SCHWERIN MBH & CO. KG	Germany	local	Eurawasser	WWW	91,264
MIDEWA WASSERVERSORGUNGSGESELLSCHAFT IN MITTELDEUTSCHLAND MBH	Germany	local	Veolia	WWW	323,000
THESSALONIKI WATER SUPPLY AND SEWERAGE CO.	Greece	local	listed company	WWW	1,200,000
SZEGEDI VIZMU ZARTKORUEN MUKODO RESZVENYTARSASAG	Hungary	national	Veolia	WWW	200,000
ACEA ATO 2—GRUPPO ACEA—SOCIETA' PER AZIONI IN FORMA ABBREVIAT A ACEA ATO 2 S.P.A.	Italy	national	ACEA	WWW	3,705,995
PUBLIACQUA S.P.A.	Italy	national	ACEA	WWW	1,300,000
ACQUEDOTTO DEL FIORA SOCIETA' PER AZIONI O IN FORMA ABBREVIATA AC QUEDOTTO DEL FIORA S.P.A.	Italy	national	ACEA	WWW	392,990
UMBRA ACQUE S.P.A.	Italy	national	ACEA	WWW	501,186
NUOVE ACQUE SPA	Italy	national	ACEA-SUEZ	WWW	267,244
INDAQUA MATOSINHOS—GESTAO DE AGUAS DE MATOSINHOS. S.A.	Portugal	national	ANTIN INFRASTRUCTURES SERVICES	WWW	170,000

(continued)

Table C.1 (continued)

Company	Country	Regulatory model	Main shareholder	Segment	Inhabitants
ADG—AGUAS DE GONDOMAR. S.A.	Portugal	national	ANTIN INFRASTRUCTURES SERVICES	WWW	150,000
INDAQUA FEIRA—INDUSTRIA DE AGUAS DE SANTA MARIA DA FEIRA. S.A.	Portugal	national	ANTIN INFRASTRUCTURES SERVICES	WWW	115,000
AGUAS DO PLANALTO—SOCIEDADE CONCESSIONARIA DO SISTEMA DE ABASTECIMENTO E DISTRIBUICAO DE AGUA. S.A.	Portugal	national	Saur	W	71,113
ADC—AGUAS DE CASCAIS. S.A.	Portugal	national	Saur	WWW	214,000
AGUAS Y SERVICIOS DE LA COSTA TROPICAL DE GRANADA AIE	Spain	local	Acciona	WWW	220,000
AGUAS DE VALENCIA SA	Spain	local	Eugenio Calabuig Gimeno	W	3,000,000
AQUONA GESTION DE AGUAS DE CASTILLA SAU	Spain	local	Veolia	W	1,100,000
AIGUES DE BARCELONA. EMPRESA METROPOLITANA DE GESTIO DEL CICLE INTEGRAL DE L'AIGUA SA	Spain	local	Veolia	WWW	3,345,695
SOCIEDAD GENERAL DE AGUAS DE BARCELONA SAU	Spain	local	Veolia	WWW	3,000,000
VIAQUA GESTION INTEGRAL DE AGUAS DE GALICIA SA	Spain	local	Veolia	WWW	765,000
YORKSHIRE WATER SERVICES LIMITED	United kingdom	national	citigroup	WWW	5,700,000
SEVERN TRENT WATER LIMITED	United kingdom	national	listed company	WWW	8,000,000
UNITED UTILITIES PLC	United kingdom	national	listed company	WWW	7,300,000
ANGLIAN WATER SERVICES LIMITED	United kingdom	national	listed company	WWW	6,000,000

Table C.2 The dataset

Name	Description	Measure	UoM
Website	company or group website	N/A	N/A
Main shareholder	Main investor who controls the company	N/A	N/A
Segment	Water. or Water and Wastewater	N/A	N/A
Inhab	Number of inhabitants served collected from website in 2025	N/A	Number
WEC	Proxy of Water Expenditure per Capita	Value of Production/Inhabitants	Euro
VoP	Value of production	Revenues +/− Stocks variations of raw materials and goods	Euros
P/L	Profit (+) or loss (−)	Total revenues – Total costs	Euro
ROA	Return on Asset	$net\,income / total\ asset \times 100$	%
ROE	Return on Equity	$net\,income / equity \times 100$	%
ROCE	Return on Capital Employed	$net\,income / (total\ asset - current\ liabilities) \times 100$	%
TA	Total Asset	Sum of assets	Euro
EBITDA margin	Incidence of EBITDA (earning before interests. Taxes. depreciations and amortizations) on the Value of Production	$EBITDA / VoP \times 100$	%
EBIT margin	Incidence of EBIT (earning before interest. Taxes. depreciations and amortizations) on the Value of Production	$EBIT / VoP \times 100$	%
PL margin	Incidence of Profit or Loss on the Value of Production	$Profit(or\ Loss) / VoP \times 100$	%
Gearing	Incidence of net financial debt on shareholders' equity	$(Debt - Cash) / Equity \times 100$	%
Turnover ratio	Incidence of the Value of Production on Total Asset	$VoP / Total\ Asset$	Number
Staff	Number of employees	N/A	Number
INV	Investments	$TA_a - TA_{(a-1)} + DA_{(a)}$	Euros
VoPpc	Value of Production per capita	$VoP / inhab$	Euro
TApc	Total Asset per Inhabitant	$Total\ Asset / Inhab$	Euro

(continued)

Table C.2 (continued)

Name	Description	Measure	UoM
Invpc	Investments per inhabitant	$INV/Inhab$	Euros
OPEXpc	OPEX per inhabitant	$OPEX/Inhab$	Euros
DApc	Depreciation and Amortization per inhabitant	$DA/Inhab$	Euros
EBITpc	EBIT per inhabitant	$EBIT/Inhab$	Euros
Av. VoPpc	Average of the inxed on the period observed (2018-2023)	$\sum_{a=2018}^{2023} VoPpc/6$	Euro
Av. TApc	Average of the index on the period observed (2018-2023)	$\sum_{a=2018}^{2023} TApc/6$	Euro
Av. ROA	Average of the index on the period observed (2018-2023)	$\sum_{a=2018}^{2023} ROA/6$	%
Av. ROE	Average of the index on the period observed (2018-2023)	$\sum_{a=2018}^{2023} ROE/6$	%
Av. ROCE	Average of the index on the period observed (2018-2023)	$\sum_{a=2018}^{2023} ROCE/6$	%
Av. EBIT margin	Average of the index on the period observed (2018-2023)	$\sum_{a=2018}^{2023} EBIT\%/6$	%
Av. EBITDA margin	Average of the index on the period observed (2018-2023)	$\sum_{a=2018}^{2023} EBITDA\%/6$	%
Av. PL margin	Average of the inxed on the period observed (2018-2023)	$\sum_{a=2018}^{2023} PL\%/6$	%
Av. Turnover ratio	Average of the inxed on the period observed (2018-2023)	$\sum_{a=2018}^{2023} Turnover\ ratio/6$	
Av. Gearing	Average of the inxed on the period observed (2018-2023)	$\sum_{a=2018}^{2023} Gearing/6$	%